A Practical Guide to Exibiting Poultry Eggs

Haeley Reeves

A Practical Guide to Exibiting Poultry Eggs © Haeley Reeves 2026

ISBN: 978-1-923746-54-1 (paperback)

Published in Australia by Haeley Reeves and InHouse Publishing.
www.inhousepublishing.com.au

IN HOUSE PTY LTD PUBLISHING

NATIONAL LIBRARY

A catalogue record for this book is available from the National Library of Australia

Dedication

I could not have completed this book without the support and encouragement of my number one egg, A.K.A husband, Paul. For the hours of listening to me talk about eggs, helping by creating graphics and editing photos, and, finally, making me judge accessories. Thank you.

Acknowledgements

Ross Coe for reading and re-reading my drafts and sharing his insights from years of teaching.

Megg Miller for her guidance, expertise and generosity in sharing her extensive knowledge of all things eggs and poultry.

Doug Somerville for all his insights into pheasants and for painstakingly noting the size and weights for the eggs of different breeds, as well as sending me some terrific pictures.

Melissa Hall of Ambleside Poultry Photography for kindly sharing her amazing pictures.

David Simons for allowing me to steward the eggs at Moreton Bay Poultry Club and for his support and encouragement with this book.

Kevin Smith for giving me to opportunity to judge eggs as well as encouragement and support with this book.

Clare Beebe for her generosity in sharing how judging and exhibiting are done in the UK.

All those who asked me questions about exhibiting eggs enabled me to answer those common questions in this book.

Anyone else I forgot to mention, thank you.

Contents

Introduction

Dear Reader,

I wrote this book as I could not find one written specifically for Australian standards or with the practical details I wanted. I have tried to explain in plain language how to exhibit eggs with all the practical considerations required to show the best egg. I hope you find this book helpful and will feel confident to enter eggs for exhibition.

For poultry clubs, I have included details on running egg classes. For anyone interested in how to judge eggs, I describe all the tools and methods used to select the winning eggs.

I hope this book encourages both the poultry keeper to exhibit eggs and poultry clubs to offer egg classes at their annual shows.

I would love to see your entries, and you can share your pictures on the Facebook Page, Exhibiting Eggs.
https://www.facebook.com/exhibiteggs

If you are interested in learning more about nature's amazing eggs, there is a list of books at the end.

Wishing you successful Exhibiting,

Haeley

Exhibitor

THIRD PLACE
SECOND PLACE
FIRST PLACE
CHAMPION PLATE OF EGGS
RESERVE CHAMPION PLATE OF EGGS

Egg Exhibitor

What is an egg exhibitor?

An egg exhibitor is a person/partnership/family that owns a fowl (bird) that has laid eggs that have been entered into an agricultural/poultry competition (show) for the purpose of having them judged to a described standard. Note that you do not have to have bred the bird that laid the egg.

This exhibitor section of the book is for those new to exhibiting eggs in agricultural or poultry competitions. The intention is to cover the basics, so you, the reader, know exactly what is required to select, prepare, and enter eggs in competition.

Creation of the Egg

Why do you need to know how an egg is created?

If you understand the stages of the creation of an egg, it will help you to identify problems with eggs laid, and you can then research ways to correct any issues your hens may encounter. Note that a hen's reproductive system contains one ovary, which is on the left-hand side. It is believed that the right ovary disappeared during the evolution of birds in order to reduce weight to assist birds with flying. If the single ovary fails, there is no backup ovary, and the hen will be infertile and may not even lay.

For chickens, the time between eggs is more than 24 hours. Consequently, a hen will lay later and later each successive day and then appear to skip a day.

Times given are approximate. Variation between breeds can be expected. Get to know your bird's cycle. Noting this pattern can help identify the optimum time to collect eggs to exhibit. By removing eggs from the nest as close to laying as possible, it almost guarantees clean, unspoilt eggs.

Species	Approx. Time Between Eggs Laid
Chicken	24-26 hours
Duck	23-26 hours
Goose	26-48 hours
Guinea Fowl	24 hours
Pea Fowl	48 hours
Quail	1-4 eggs in 24 hours at peak
Turkey	24-32 hours

Stages of Egg Creation
Ovary Where immature oocytes (eggs) are stored and develop. As the oocytes mature, they are released from the stigma of the ovary.
Infundibulum Catches the egg as it is released from the ovary. Here is where fertilisation occurs if a cock bird is present in the pen.
Magnum Inner and outer shell membranes are added, in addition to some water and mineral salts.
Ithmus Albumen is secreted and layered around the yolk.
Uterus (shell gland) Here, water is added, making the thinner outer albumen. Next, the shell (calcium carbonate) is added. For dark laying breeds, the shell pigment is added.
Vagina and Cloaca The last section to pass through before laying. Here, the egg rotates and comes out blunt end first. Contains sperm storage tubules.
Total Time 24-26 hours (breed dependent)

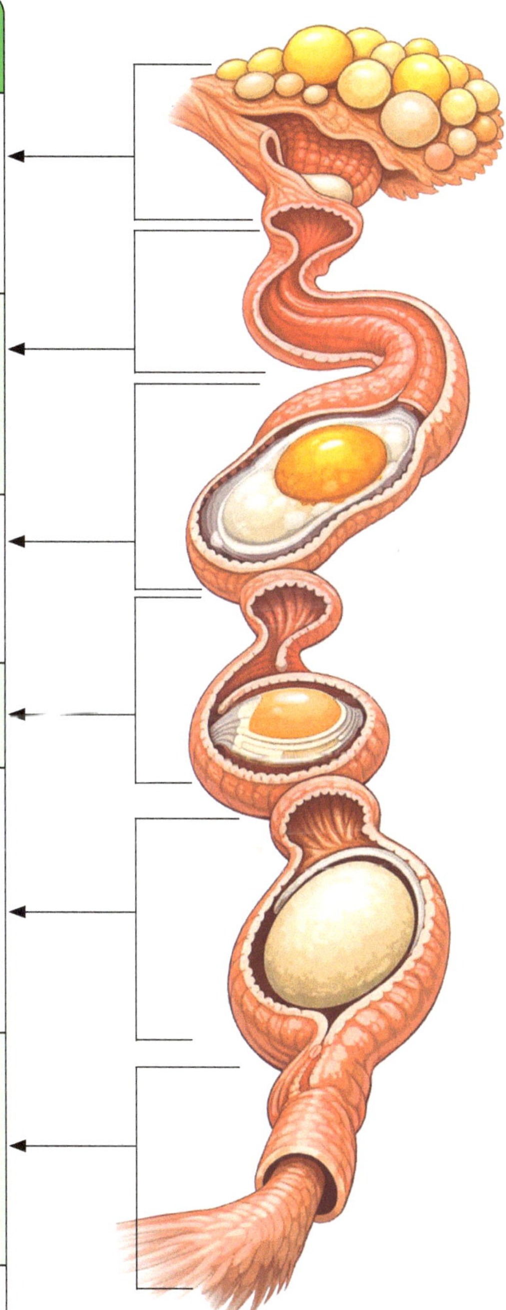

Oviduct — internal reproductive parts of the female

Parts of an Egg (Nomenclature)

Judging Order	Name	Description
1	Cuticula	Also known as the bloom on the outside of the shell. The cuticula is a thin membrane/ coating that helps to keep out dust and bacteria.
2	Shell	Made almost entirely of calcium carbonate ($CaCo_3$) crystals. It is a semi-permeable membrane, allowing air and moisture to pass through.
3	Vitelline Membrane	Forms the outer membrane of the yolk and is covered by the chalaziferous layer.
4	Air Sac	Small air space when freshly laid, which increases in size as the egg ages.
5	Albumen	Also known as egg white, the albumen accounts for 66% of the liquid weight of an egg. It consists of several alternate layers, the thin outer albumen and the thick inner albumen. The albumen provides an anti-bacterial barrier for the yolk and developing embryo.
6	Chalazae	Chalazae are twisted membranous strands that join to either side of the yolk, keeping the yolk centred in the middle of the albumen.
7	Chalaziferous (thin) Layer	This fibrous layer of albumen directly covers the entire yolk and connects to the chalazae strands.
8	Yolk	The yolk is a sphere of nutrient-dense emulsion providing all the requirements for a developing embryo.
9	Blastoderm	The blastoderm is a germinal disk of cells on the top surface of the yolk.

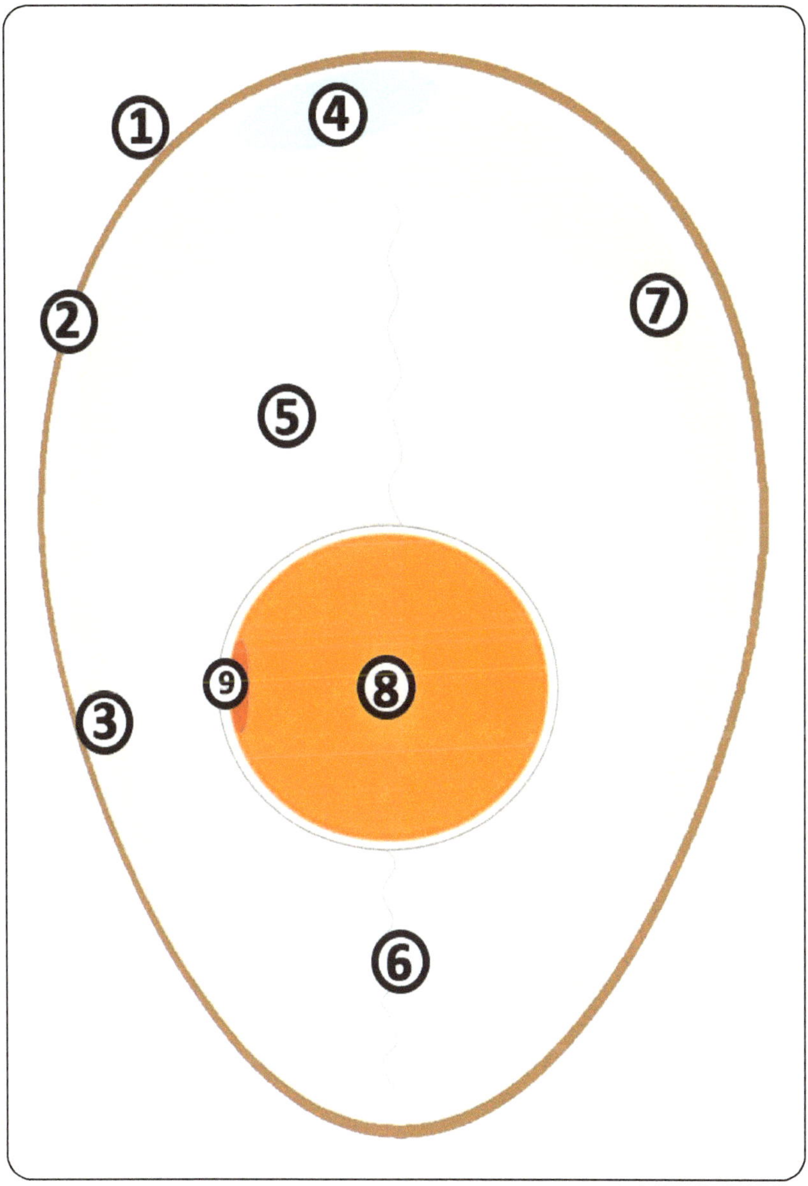
1
2
3
4
5
6
7
8
9

Characteristics of an Egg

External

Shape

Why is the ideal shape so important?
It produces optimum success at hatching for the chick and provides sufficient room for growing. An ideal egg is wide at one end for the air sac where the chick takes its first breath before breaking out of the shell. The other end is pointed to help keep the egg in the nest when it is rolled back and forth by the hen during incubation.

Ideal Egg Shape

Chick about to hatch at 21 days

The following shapes and their descriptions relate to chicken eggs in general. Please note that there are variations in the shape of the egg for different species. These are described in more detail in the species-specific section.

Poor Egg Shapes

Elliptical
No pointed end
Narrow, flat-sided

Spherical
Round like a ball

Oval
Rounded at both ends

Biconical
Pointed at both Ends

Conical
Very pointed at one end
Elongated in the middle

Ridged/
Wrinkled/Corrugated
Uneven shell

Body Checked
Broken and mended
(note the ridge)

Truncated/Slab-Sided
Flat on one side, rounded on the other.

Dwarf/Fairy/Wind
Very small and often has no yolk

Soft-Shelled/Shell-Less
No shell
Membrane Only

Size

The size of the egg is proportional to the size of the bird. A larger egg results in a larger chick, which grows into a bigger bird, producing bigger eggs.

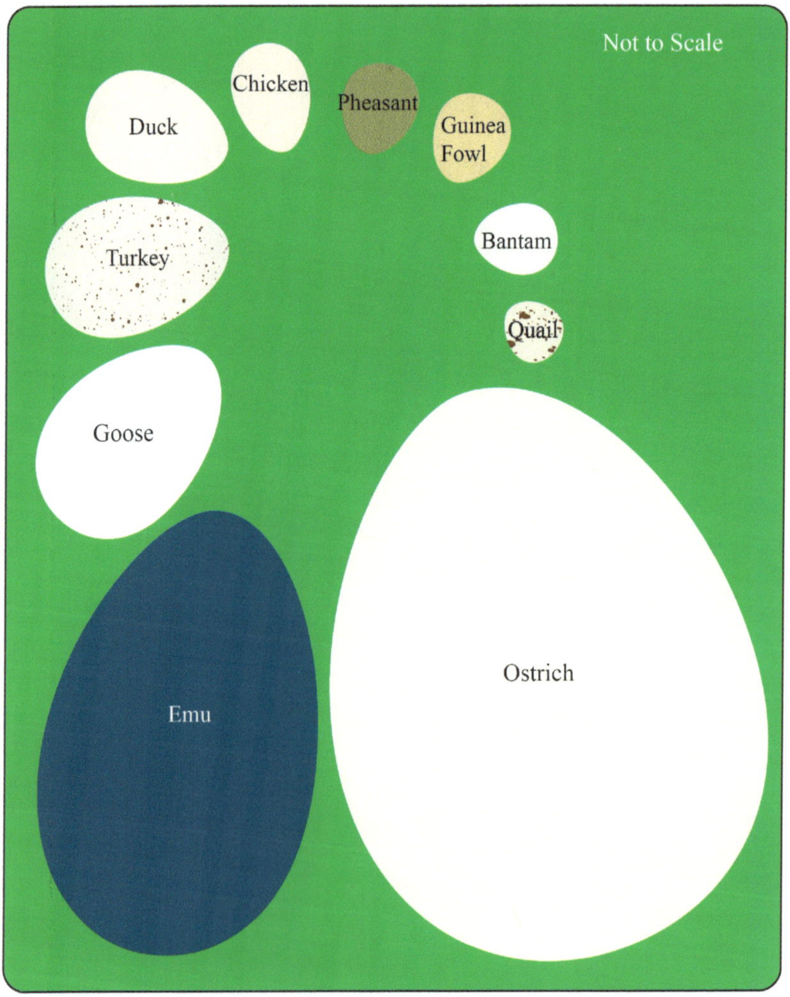

Weight

Why is weight important?

For breeding purposes, the weight of the egg is in proportion to the size of the chick that will hatch. Weight is a factor in hatchability. Medium-sized eggs have the highest hatchability, while very large or very small eggs have lower rates.

In the commercial world, consistent egg size is sought, and the eggs are graded according to size. The weight of the egg is also used in recipes. The household cook needs to know the size (weight) of the egg purchased which is why cartons are clearly labelled.

In competition, eggs are grouped into the same weight class for better comparison when judging. Judges also look for a consistent weight of eggs on a plate.

Here is a guide for egg weights according to the Australian, New Zealand, and British standards.

- Chicken bantam — less than 42.5g
- Chicken large — 49.6 pullet, 56.7g average
- Duck bantam — less than 63.8g
- Duck large — 70.9-92.2g
- Turkey — 70.9-92.2g
- Goose light — from 141.8g
- Goose heavy — up to 198.6g

Guide to weight for species not in the current standards

- Quail — average 10g
- Pheasant — average 20g-35g
- Guinea Fowl — average 46g

<table>
<tr><th colspan="2">Commercial Sizes for reference ONLY
NOT used for EXHIBITING</th></tr>
<tr><th>Australian Commercial</th><th>UK Commercial</th></tr>
<tr><td>Jumbo: 68g
Extra large: 60g
Large: 52g
Medium: Around 43g</td><td>XL (Very large, 73g+)
L (Large, 63-73g)
M (Medium, 53-63g)
S (Small, <53g)</td></tr>
<tr><td colspan="2">Canadian and American Chicken Egg Sizes
These are the same definitions used for commercial grading of chicken eggs in Canada.</td></tr>
<tr><td colspan="2">Jumbo — more than 70g
Extra large — between 63-69g
Large — between 56-62g
Medium — between 49-55g
Small — between 42-48g
Peewee — less than 42g</td></tr>
</table>

I recommend that you refer to the egg standard and schedule when entering your eggs. If there is no standard available, check the terms and conditions in the show schedule.

If you are still unclear, seek clarification from the chief show steward. Better to ask when you are unsure, as entering eggs into the wrong class may result in being passed over (not assessed) or disqualification. When pullets (hens under 12 months old) first lay at around six months of age, the eggs will be small. Eggs will increase in size as the hen ages. There is also a noticeable increase in egg size after the hen's first moult.

Be mindful of this when entering eggs some time before a show, particularly regarding pullet eggs. You may find your eggs are overweight when entered into a bantam class or potentially underweight for large classes.

For more accurate measurement of weight, use a set of scales capable of measuring less than a gram. This will help you to ensure that your eggs are not over- or under-weight for a class. This also helps you select eggs that are close in weight, ideally choosing eggs that weigh the same and are similar in appearance.

NOTE: Keeping your eggs in the fridge will help to maintain the humidity and reduce the loss of weight prior to the show.

Analogue scales are not as accurate as digital scales, are also expensive and hard to find, and are usually imperial rather than metric.

Digital scales are cheaper, readily available, more accurate, and much quicker and easier to use.

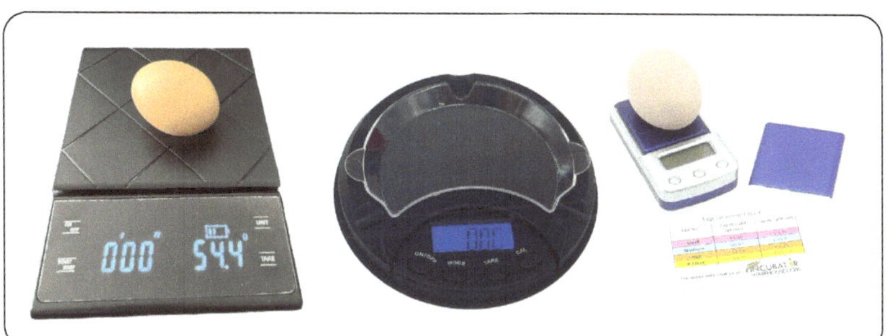

Shell Texture

Good Shell Texture - is smooth, even texture free from porosity (open, grainy look and feel).

NOTE: When selecting eggs to exhibit in competition, feel all over the egg to check for even texture, particularly the ends of the egg. Speckled eggs with larger spots tend to have a raised surface, as the pigmentation is thicker, and this is considered normal.

Calcification

Extreme spots like this would be considered a fault

Avoid selecting a shell with calcification for the exhibition plate.

Bad Shell Texture - can be a chalky and or gritty surface. The surface of dark brown eggs is more prone to being scratched in the nesting box. Collect eggs as soon after laying as possible to prevent the spoiling of the cuticle.

The bullseye pattern occurs when two eggs collide in the oviduct.

In this photo, the white is a coating of calcium that has built up due to the delay in laying.

Bullseye pattern

Shell mottling is where there are more translucent areas on the shell, which can be seen very clearly when candling. Some breeds are more prone to mottling, as are large eggs and those laid by older hens. The more mottling, the more the egg is prone to losing moisture. This results in quicker aging of the egg.

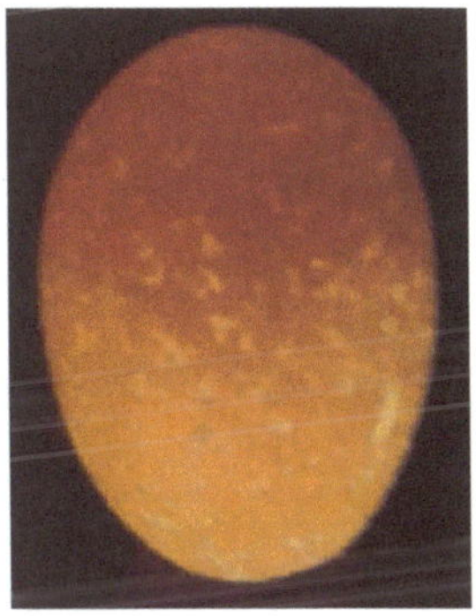

Cracked Shell

Fine cracks can be hard to detect in normal daylight. Using candling (a light is shone through the egg), will show cracks very clearly. Candling is used extensively in commercial environments, as cracks make the egg very susceptible to infection with bacteria. These eggs are NOT suitable for human consumption.

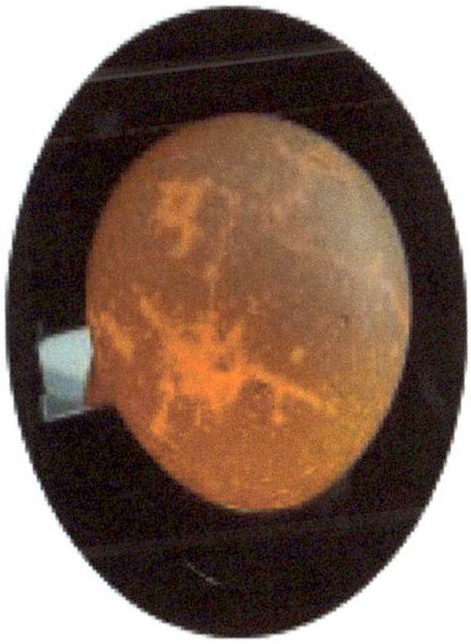

There should be no need to polish or oil eggs. Clear evidence of tampering with the surface of the egg will result in disqualification.

Soft-Shelled/Shell-Less

Eggs laid without shells are not suitable for exhibiting or hatching. Note how this egg has lost shape just one day after laying due to a significant loss of moisture.

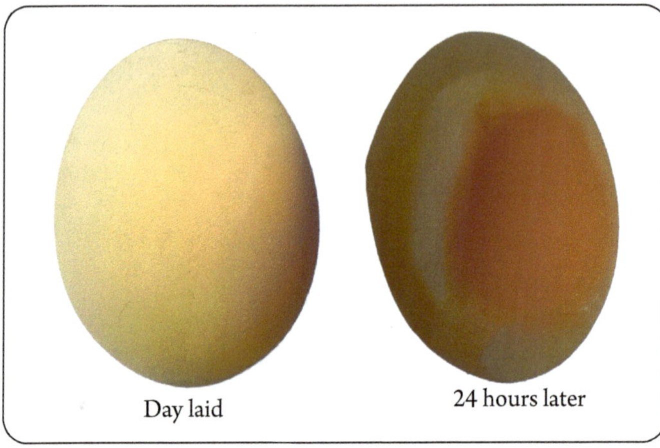

Shell Colour

The terminology of egg colour varies considerably before the writing of a standard. Therefore, it is important to become familiar with the terms used in the standard.

The shade of colour may vary, so grouping of eggs is made to keep the number of classes smaller.

It is important, regardless of the colour, to have uniformity of shade across all eggs on a plate.

Note that some colour groups, like tinted and cream, can be a point of discussion and disagreeing opinions. It is important to clarify if there is a specific mention of colour definitions in the show schedule.

Here is an example from the Sydney Royal Easter Show, where tinted is clearly defined as follows:

> *"Tinted eggs are to be very pale brown to cream, including light beige, ivory, or off-white. Tinted does not include white, medium or dark brown, blue, or green."*

The colour of the shell in classes is not limited to those in the standard. If in doubt on which class to enter eggs, check with the show steward.

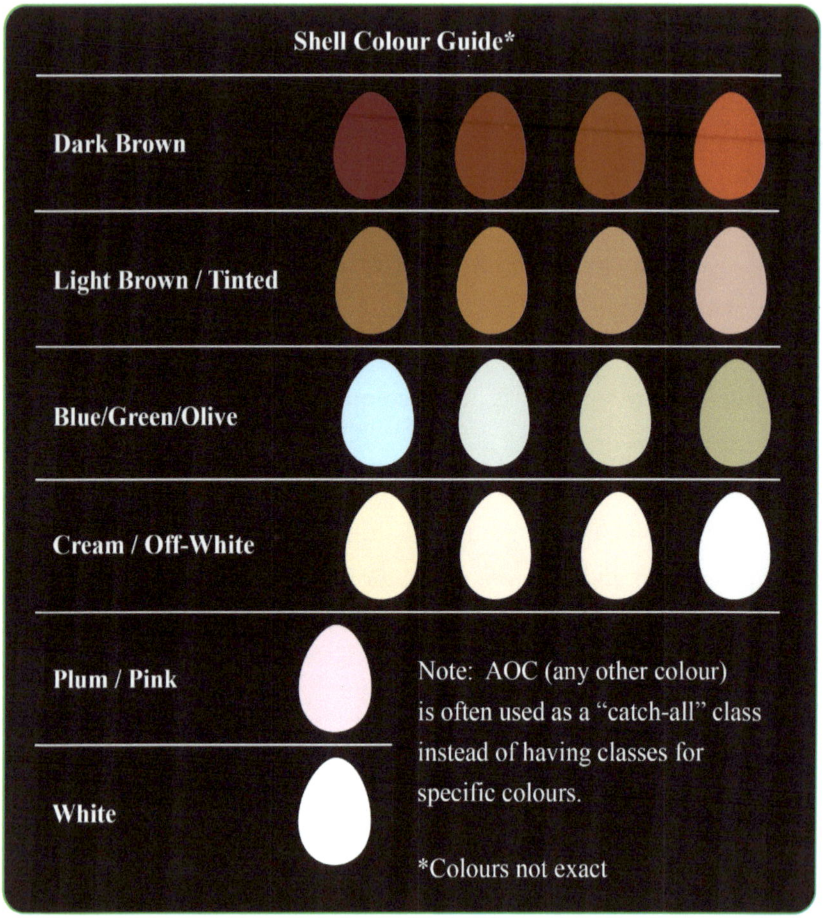

Examples of Colour Classes

Mottled and Speckled Eggs

These should be grouped in the class based on the ground colour, unless there is a breed-specific class, such as Welsummer. Eggs with mottling and speckles that form a regular pattern over the entire surface are preferred.

Warning: DO NOT "enhance" the colour of the egg for standard classes.

The steward may report this behaviour to the local poultry exhibition organisation in the jurisdiction in which the show is held. In addition, your entry can be disqualified or "passed over".

Internal

The criteria for measuring the internal contents of an egg are divided into its component parts: **yolk**, **albumen**, and **chalazae.** Optimum characteristics are described for each as follows.

Yolk Size

The size of the yolk is affected by the age and the breed of the bird. Duck, turkey, and goose eggs have a much larger yolk by proportion compared to chicken eggs. The larger the yolk, the larger the egg.

Here are examples showing the albumen and yolk size.

Goose	Guinea Fowl	Chicken
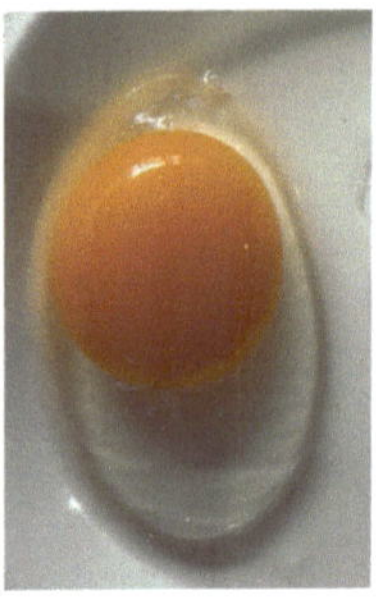	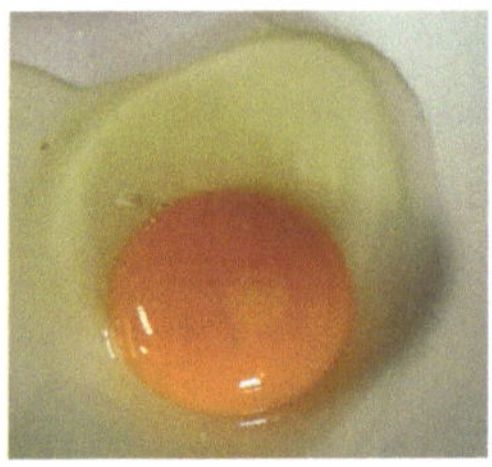	

Yolk Colour

Egg yolk colour is not dependent on breed, but on diet.

Rich golden-yellow is deemed to indicate the optimum nutritional content, and this is reflected in the egg standard. Natural-source pigments in diets are derived from marigold petals or red peppers.

Red/Orange	Lots of green food, especially grass. The darker the colour of the yolk, the more carotenoid-rich the plant, the more orange the yolk
Yellow	Lots of corn and greens
Light Yellow	High percentage of barley and wheat

Albumen

The contents should be free of blood, meat spots, or cloudiness or any tint of colour. It should contain two clearly defined layers of albumen.

NOTE: Too much green food can cause a greenish tinge in the albumen. Check the egg yolk colour several weeks before collecting for a show, giving you time to adjust the food if required.

Chalazae

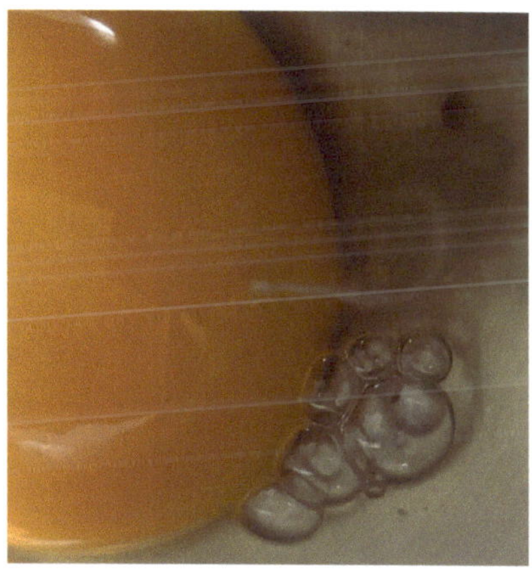

Two thick white cords are attached to either side of the yolk. The chalazae's purpose is to keep the yolk in the centre of the albumen as the egg rotates during incubation.

For exhibiting purposes, avoid spinning the egg or shaking vigorously, as this can detach the chalazae.

Matching and Uniformity

Matching and uniformity are where the eggs in an exhibit/plate are assessed for being the same weight, size, shape, shell texture, and colour. For speckled, mottled, or blotched eggs, a similar pattern for each egg is preferred.

NOTE: Selecting eggs from the same bird will assist with uniformity.

To select your exhibit, line up all the eggs you have available and compare their size and shape. Use eggs that match as closely as possible.

In this example, three eggs match very closely and were most likely laid by the same bird.

Measuring Freshness

External Checks

Cuticle/Bloom

Examine the cuticle closely, checking that the bloom is without dull or stale appearance. Lighting affects the appearance of the shell. Daylight is preferable; otherwise, try to work under a good level of lighting, preferably with a daylight bulb.

Rotten egg

The shell of this white egg is clearly discoloured from the inside and indicates that this egg is old and rotten inside. This egg would be passed over in a judging situation.

Air Sac

Why is the size of the air sac an important measure of freshness? From the moment the egg is laid, it loses water by evaporation through the shell pores. There is no air sac at the moment of laying. The contents contract slightly as the egg cools, and the air sac then starts to form.

The rate of water loss is determined by the temperature, relative humidity, and the air flow around the egg. The quality of the egg deteriorates as the water in the contents evaporates through the shell. The smaller the air sac, the fresher the egg.

Position of the air sac?

For purposes of hatchability, the air sac should be at the broad end of the egg to one side. This enables the chick to bring up its head and reach to pierce the membrane of the air sac. The chick pierces the air sac to take its first breath and exercise the lungs, before pipping the shell. An air sac in the centre is a disadvantage to the chick as they are not able to get their head back far enough to reach up. Therefore, a centred air sac is considered a fault.

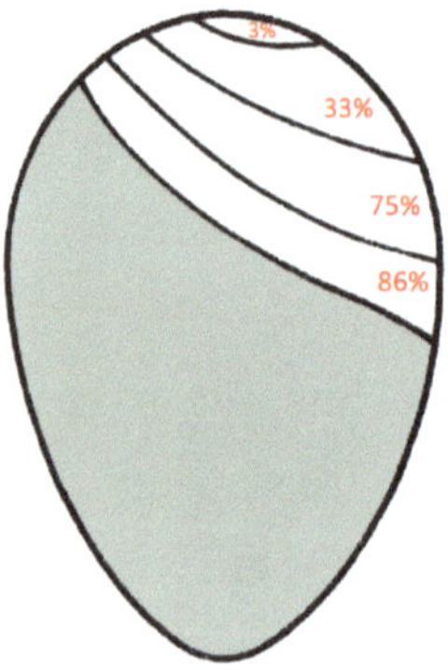

The percentages shown are the loss in weight of the egg from the time of laying. The percentage loss will be affected by many factors, depending on how the egg has been stored. Regardless of species, the air sac follows the same pattern of enlargement over time.

In large chicken breeds, the air sac is no bigger than 1.5cm in diameter and 1cm in bantam eggs. The air sac is of the same proportion for the larger size eggs laid by ducks, turkeys, and geese.

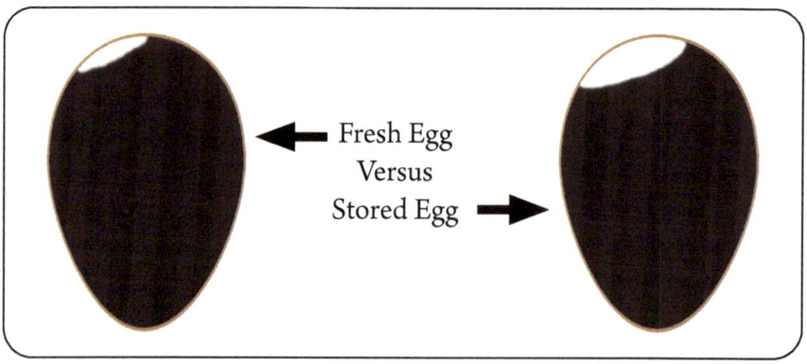

When candling an egg, rotate the egg to see if the air sac is off to one side.

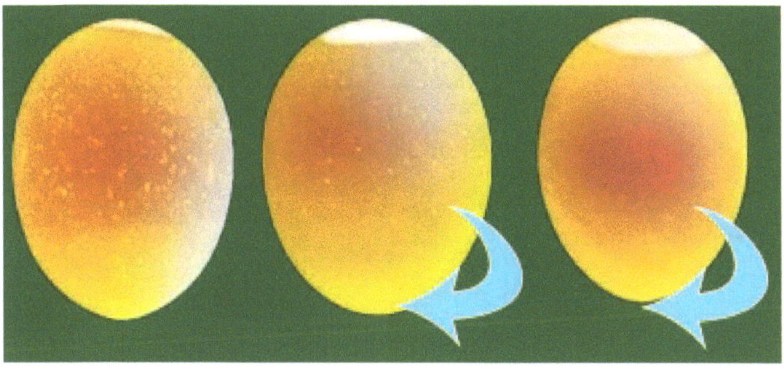

Air Sac Faults

Checking for faults is made easier when candling.
Seeing the size and shape of the air sac enables you to discard those eggs with faults.

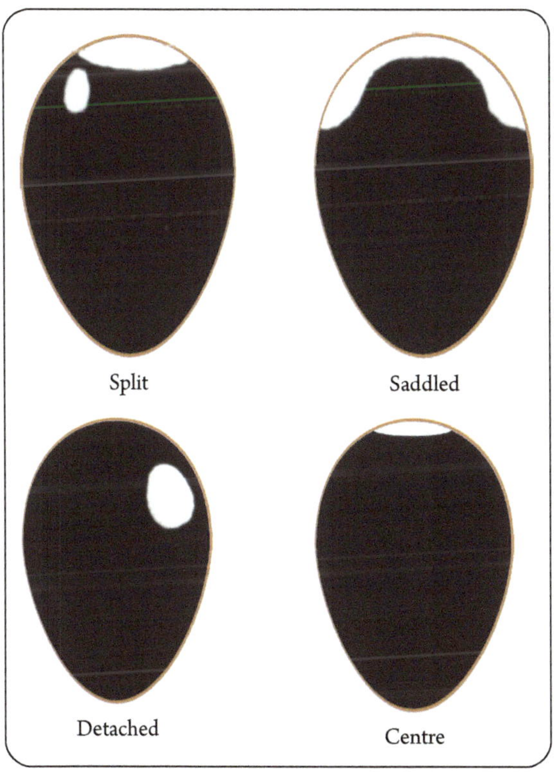

Candling

Candling is the only way to see the size and position of the air sac without cracking the egg.

Use an OvaScope or a high-powered torch to see the contents of the egg. The modern LED light bulbs also have the advantage of providing light without heat.

The higher the lumens, the more illumination is achieved, and this is most helpful in order to get a clearer image of the air sac when candling dark brown eggs.

A more powerful torch will make it easier to candle the egg.

NOTE: Dark brown eggs are much harder to candle than white, as the dark cuticula obscures the inside of the shell.

Float Test

Floating in water is a well-known method. The egg that floats is old, has lost water, and absorbed enough air for the egg to float to the top. A fresh egg will rest on the bottom of the container.

This method can easily be done at home, but it is not practical for assessment in competition.

Internal Checks

Air Sac

After cracking the egg, look at the blunt end. Inside the shell will be the air sac. It should be on one side. The size of the air sac will be in proportion to the freshness of the egg. A fresh air sac will be small and the membrane taut; whereas an old egg will have a much bigger membrane visible and will be wrinkled.

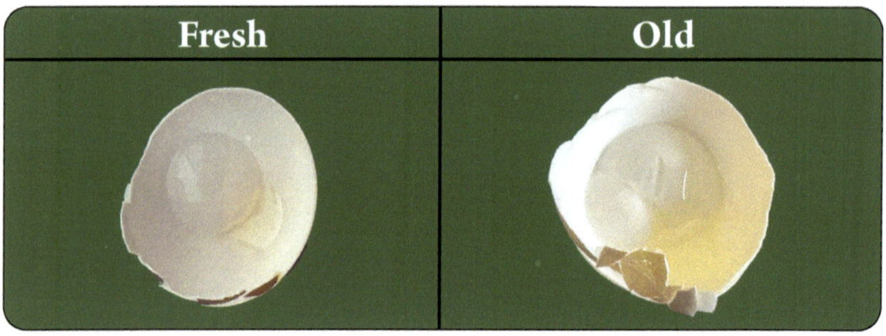

Yolk

The yolk should have a smooth top surface and be raised above the height of the albumen. A wrinkled yolk is a sign that the vitelline membrane surrounding the yolk has started to lose strength and viscosity, indicating that the egg is not at its freshest.

Mottled yolk

The yolk should be of an even yellow colour, not mottled. Mottling can be an indicator of an older egg. It can also be a sign of an inconsistent diet.

The size of the yolk is affected by the age and the breed of the bird. Duck, turkey, and goose eggs have a much larger yolk by proportion compared to chicken eggs.

Abnormal yolk shapes are the result of variations in the strength of the vitelline membrane that surrounds the yolk. Weaker parts of the membrane allow the yolk to bulge out from the corners to form a square. These formations are very rare and can be heart-shaped, U-shaped, and triangular.

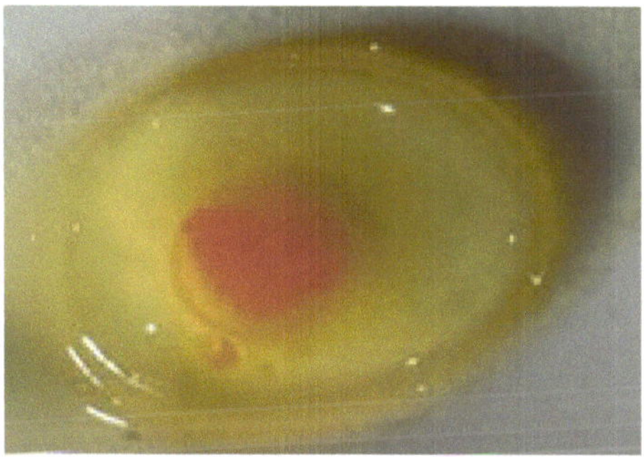

This yolk is an example of the triangular shape. Here, some of the yolk has detached and formed a blob shape. (The yolk was **not** broken when being cracked open by the judge at the show where the author was stewarding).

Blood on yolk occurs with young pullets as they come into lay. Blood attaches to the yolk from the torn stigma of the ovary and remains attached as the egg moves down the oviduct. Blood spots in the albumen are considered a fault. These eggs should be discarded.

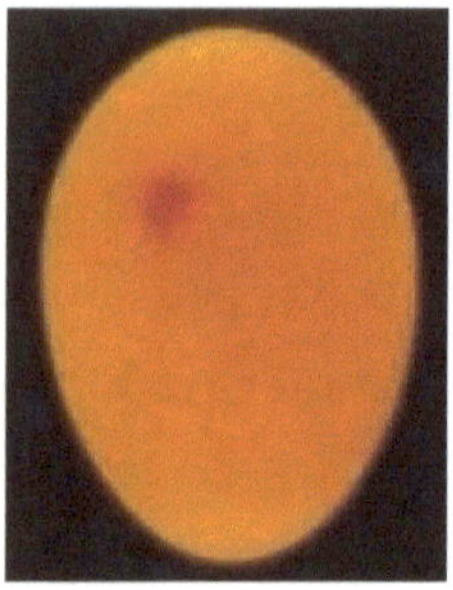

It is not always possible, particularly for brown-shelled eggs, to identify blood or meat spots via candling. This example shows a candled white-shelled duck egg with a blood/meat spot.

Albumen

Albumen is stale if it lacks differentiation (two distinct layers) and is watery and runny.

Cloudiness of the albumen is due to the natural presence of carbon dioxide that has not had time to escape through the shell. This is an indication of a very fresh egg. As an egg ages, the carbon dioxide escapes, and the egg white becomes more transparent.

Waterfowl albumen must have clear albumen. This is why it is preferred that they are in a class of their own.

As mentioned, meat spots are a small piece of tissue from the wall of the oviduct that attaches to the albumen. Like blood spots, they are difficult to detect by candling.

Select competition eggs from a hen with a history of laying eggs without meat spots. This means you are more likely to reduce the chances of a meat spot occurring. This applies to both choosing eggs for competition and for incubation.

Factors that Affect Egg Quality

This section covers the basic factors to consider to optimise the quality of the eggs you exhibit. This is not an exhaustive list, and I encourage you to undertake further research. At the back of this book are some reading resources that you may find of interest.

Genetics

- Do not breed from bad eggs. Find your consistently good laying hens, set only the best eggs laid (good size for breed, ideal shape, consistent colour, good shell).
- Breeds with yellow pigment in their legs and those hens with paler legs have been found to lay more.

Food and Nutrition

- Carotenoids are crucial in poultry nutrition for natural pigmentation (yolk, skin, legs, beak) and health benefits, acting as antioxidants to reduce stress, boost immunity, and improve bird performance. In commercially produced feeds, carotenoids are added to meet consumer demand for vibrant colours of yolks and improve bird wellness. Commercial carotenoids are made from natural sources like plants (carrots, marigolds, tomatoes, algae). Check your poultry food for their presence.
- Use plenty of fresh, clean water daily — 0.2 litre per bird per day average. In hot weather, more water is required.
- Provide a balance of quality ingredients.
- Pellets and mash provide more consistent nutrition compared to mixed grain, as birds will pick out their favourite seed, e.g. eat sunflower and leave sorghum.
- Chickens require methionine, an essential amino acid, for growth and development, and can obtain it from both

natural and synthetic sources. Natural sources include fish meal, earthworms, insects, and certain plant proteins, while synthetic methionine (DL-methionine) is also used in poultry feed.

- Linoleic acid stimulates laying and helps increase the size of the egg.

> **TIP:** Adding a small amount of vegetable oil to the feed, just enough to coat the pellet/grain (do not saturate), will stimulate laying.

- B vitamins are crucial for chickens, playing a vital role in energy conversion, growth, egg production, and overall health. Specific vitamins like B12 are important for nerve health and red blood cell formation.
- DON'T overfeed your birds — they lay less and have a higher chance of laying defective shells.

A Rough Guide to Food Quantities

Chickens		
Breed	**Weight**	**Time Period**
Bantam	36g	Day
Light Breed	143g	Day
Dual-Purpose	215g	Day
Heavy Breed	286g	Day
Broiler	643g	Day

Other Species		
Species	**Weight gram**	**Time Period**
Duck	115-225g	Day
Goose	225 900g vegetation 200-300g commercial	Day
Guinea Fowl	113-127g	Day
Pheasant	130-160g	Day
Quail	14-35g	Day
Turkey	600g-1kg	Day

- Be careful with supplements. They may affect yolk colour.
- Warning: Too much green food can make the albumen have a greenish tinge.
- Too much protein can cause thin shells.
- For breeds that lay a dark brown egg, green food darkens the brown cuticula (shell colour).
- Calcium is essential for good shell quality, so make sure there is plenty of shell grit available.

- **Soluble (calcium)** grit dissolves gradually as it is soft. Whilst shell grit will aid with grinding food, its main purpose is to provide calcium.

Size and Use of Soluble Grit

Grade	Size	Use
Fine	1.2mm	chicks, small birds/quails
Medium	2-4mm	adult chickens/poultry
Coarse	Over 4mm	geese/large ducks

- Poultry do not have teeth. They grind up their food in the gizzard. Harder, rougher stones are ideal, such as **insoluble** (digestive) grit that does not dissolve and is made from crushed granite, flint, or coarse sand. Insoluble grit lasts longer and assists with a greater conversion of nutrients from food ground in the gizzard. Make sure a variety of grit sizes is available for your birds; they will select the size they require.

Housing

- Ensure sufficient nest boxes are available for the number of hens in the pen.

No Hens.	No Nest Boxes.
3-6	2
7-12	3-4
13 or more	4-7 or more

- Despite plenty of nest boxes available, hens will often have a favourite box they all prefer to lay in.

- Put nest boxes in a darker, quieter, low-traffic area of the pen. It should be 30cm to 1m off the ground, but lower than the roosting bar. This reduces eggs laid outside the box, where they are more easily spoiled.
- Clean the nest box regularly and line it with fresh softwood shavings to minimise soiling.
- Keep the housing clean and pest-free.
- Avoid overcrowding: 1.8m^2 per bird is ideal. Using movable housing helps to mitigate stress due to a smaller space, and it works if moved regularly.
- Good ventilation will minimise the effects of ammonia from droppings and reduce heat during hot weather.

Collecting

- Ideally, as soon as they are laid, but more practically, before the end of each day.
- Put collected eggs straight into an egg carton to keep them separate to minimise the chance of damage, especially scratching the cuticle.

- Sterilise used egg cartons by popping them in the microwave for a minute, or spray with Glen 20.
- Candle your eggs to check the contents for meat spots and hairline cracks.

Storage

Place eggs pointed end DOWN in the carton. This helps to keep the yolk in the centre of the albumen.

To help maintain freshness and reduce fluctuations in temperature, keep your eggs in the fridge. The fridge provides a more consistent cooler temperature and higher humidity compared to normal room temperature.

The middle of the fridge is best, avoiding the coldest and warmest shelves at the bottom and top. Do NOT keep eggs in the door. Add a dish of water to help maintain humidity.

Commercially, eggs are stored between 0°C and 1.5°C with 80-90% humidity.

Egg Selection

- Having birds of different ages may give you a greater choice when selecting your eggs.
- Collect from the same hen for better uniformity. Collect for 14 days before the show, choosing from the seven freshest first. If any of the freshest have a defect, you can then choose from your reserve eggs aged seven to 14 days old.
- Select the shape first, as it has the largest number of points.
- **NOTE**: Shell texture, colour, and freshness each have the same number of points.
- Avoid entering the same eggs for different shows unless they are on consecutive days. Even then, you should be confident; they have not been affected by transportation or the judging process.

Seasons

When planning to enter eggs into shows, the time of year is a big factor in the production of eggs.

Day length and temperature affect the frequency of laying. Hens stop laying when they moult. The seasons are different for each latitude, but as a rule, the following occurs.

Spring	As day length increases and the average temperature rises, the spring moult will begin. Laying is usually delayed until the moult is complete, which is the best time to exhibit eggs.
Summer	As days lengthen towards mid-summer, birds will lay more. If the temperature rises consistently above 30° for more than a couple of days, hens may become broody and stop laying. The summer solstice marks the longest day of the year.
Autumn	As the day shortens, so will laying decrease. Once the autumn moult begins, hens will stop laying altogether, whilst all their energy goes into feather production.
Winter	Shorter, colder days will cause birds to decrease laying, often to stop completely. Pullets that have started to lay will lay infrequently. Winter solstice marks the shortest day of the year.

Temperature

High and low temperatures significantly affect chicken hormones, primarily by activating the hypothalamic-pituitary-adrenal (HPA) axis and increasing corticosterone levels, which is the primary glucocorticoid hormone in poultry. This can lead to various physiological and behavioural changes, including reduced egg production, impaired metabolism, and increased susceptibility to diseases.

TOO HOT – above 30°. Do anything you can to keep your birds cool during hot weather. Be aware of the signs of heat stress, such as panting and holding out wings. Successive hot days and nights can lead to death. Birds cannot sweat. They pant and raise their wings, so any means to keep your birds cool should be employed.

- Soak the ground in a shady spot, so the birds can dig down and wet their legs and bodies. As the water evaporates, it cools their lower body.
- Cool drinking water will help to bring down body temperature (pop a frozen bottle of water in their drinker), and electrolytes will help to keep them hydrated.
- Insulated roofing for housing, white painted walls, and safari roofs can all help to keep birds cool.
- Add extra drinking/water stations so that hens low in the pecking order can still access water.

TOO COLD – cold weather is associated with shorter days of winter. Hens will stop laying due to moulting and reduced day length.

- Provide insulated housing.
- Artificial heat to maintain the temperature between 19-24°C.
- Artificial light to extend the day length.
- Additional feed is required to help keep warm and assist in laying.
- Birds may experience a decrease in egg production due to hormonal shifts, primarily related to thyroid hormones (T3 and T4) and corticosterone, which are influenced by temperature and play a crucial role in metabolism and stress responses.
- Ensure that openings facing winter winds are fully or partially blocked to avoid cold draughts.

Moulting

Moulting is the term that describes when birds lose all their feathers.
Laying will reduce and often cease whilst all the energy of the bird goes into growing new feathers.

Adult birds will normally moult once a year. This usually occurs in autumn when daylight hours become shorter, but may vary depending on the time of year the bird started laying.

The first adult moult will usually occur at 16–17 months, depending on the time of year the hen started laying. This needs to be considered when entering eggs to exhibit.

Age of Birds

The age of your females needs to be accounted for when planning your eggs for showing. The age at which females start to lay will vary according to their breed, the time they hatched, and the season of the year.

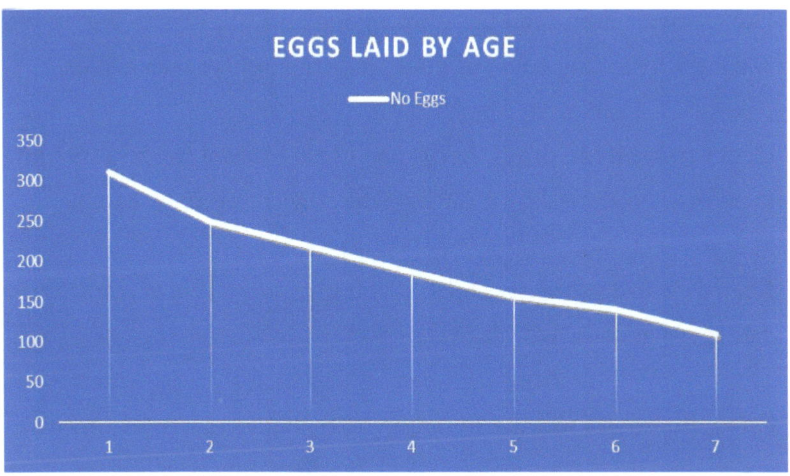

Based on average commercial chicken breeds eggs laid each year

Female birds less than 12 months old are more prone to egg faults as their reproductive system kicks into action. For example, double yolks, no shell, poor shape, and small size are very common.

Female birds over 12 months will have much larger eggs and may lay less often. The quality of the shell colour, particularly for dark brown eggs, will fade as the laying season progresses.

For all hens regardless of age, poor shell quality and egg shape may be a sign of an issue within the reproductive tract.

Stress

Stress can be caused by several factors; these are a few to consider.

- Moving birds between cages.
- Separating and joining different birds in a cage.
- Addition, changing, or removal of a cock bird.
- Changing food.
- Significant temperature changes.

Medication

Use of medications can have both good and bad effects on the bird and eggs, including conventional medicine or herbal remedies.

Check to see if the drugs/herbs you use are meant for poultry and if there is a withdrawal time during which eggs cannot be eaten.

Keep a record of the treatments you give to your birds, noting the date given and any withdrawal period.

Often, hobbyists will use off-label drugs that were manufactured for other animals. Do your homework before eating eggs if you do use these drugs.

Cleaning Eggs

Washing is allowed, however! Try to avoid doing so, if possible, as the shell is porous and the cuticula of the egg can be very susceptible to damage. Use only water and a clean, damp cloth to **very** gently remove any dirt. Minimise removing the bloom by dabbing rather than rubbing the shell.

DO NOT USE SOAP!
A round object and slippery soap
is a recipe for disaster!

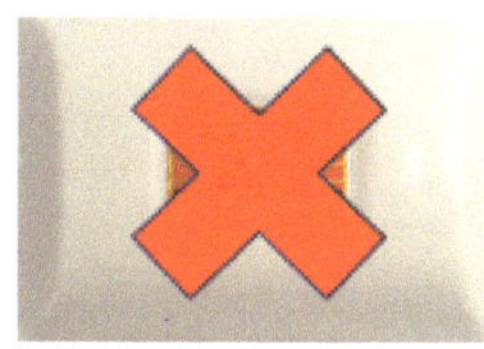

If you must wash your eggs, then avoid adding a polish by rubbing with a cloth. Instead, let the eggs air dry.

Warning! Dark brown eggs are very susceptible to marking the cuticula pigment and losing bloom, so avoid cleaning if possible.

White Eggs with a yellow stain, can be cleaned gently with Lemon Juice.

A white "magic" sponge/eraser can be used delicately to pluck dirt from the shell. It is particularly good for white eggs. However, use extreme caution for brown eggs as it can remove the brown pigment if you rub the surface, even just gently.

Transporting Eggs to the Show

After all the effort of preparing your entries, make sure you pack your eggs well for the journey to the show to ensure the eggs arrive in their best possible condition.

Small eggs — Small eggs, those from bantams, may need to be packed in a standard box with sawdust or tissues. For quail, boxes specifically for this size can be purchased.

Standard eggs — Chicken, duck, pheasant, guinea fowl.
Choose a clean egg carton. ***Label*** the outside of the carton clearly.

If reusing an existing carton, make sure it is clean. You can sterilise the carton by placing it in the microwave for a few seconds on high to kill any lurking bacteria that may spoil your eggs. Some exhibitors also spray cartons with household spray disinfectant such as Glen 20.

Wrap each egg with tissue or alternatively pack with sawdust. Place an elastic band around the egg box to keep the lid securely closed.

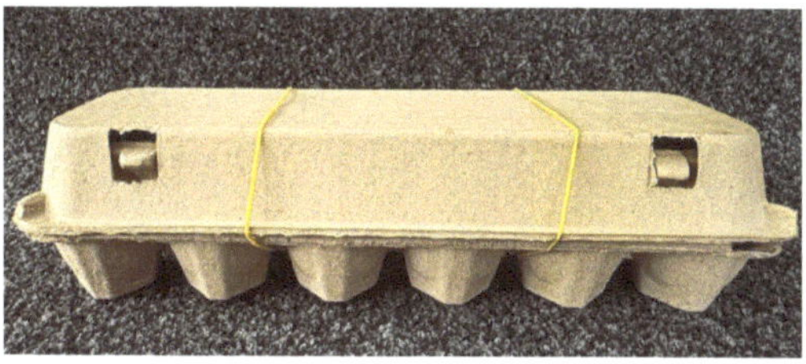

Large eggs — Duck, goose, and turkey

Sourcing cartons for these larger eggs can be a challenge. Save your 1- and 2-litre plastic milk cartons. Wash and dry the carton, then cut off the top section where the handle is.

Stuff with tissue paper or kitchen roll to hold the egg securely in place. Use a separate carton for each egg.

Following this, put them into a thick cardboard box (the thicker the better), packed to keep the cartons upright.

Put all the egg cartons together in a box for secure transport to the show, ready to hand to the steward on arrival.

Labelling

Entry No: Y649
Class: E1845
Description: Large Brown Eggs

Having multiple entries in the same container gives more opportunity for your eggs to end up on the wrong plate.

Breaking up your entries into separate, clearly labelled containers helps to mitigate this potential issue, especially in shows where the steward sets up the eggs on the plates. It will help to reduce eggs going in the wrong class or under the wrong entry in the same class.

Note: ***Do not put your name!*** It helps to keep the entries anonymous, so that no one knows who has entered eggs in a class except for the steward, until judging has finished and the awards allocated. This helps to ensure that judging is equal and fair for everyone.

Species Specifics

Egg Laying Statistics

Exactly when the hen of a species starts to lay will vary. Smaller species start earlier and lay a smaller egg. The number of eggs laid by a particular female will reduce for each successive year, regardless of species and breed.

Egg Laying Information			
Species	**Average per year**	**Age start laying**	**Season for laying**
Chicken	200-300	4-6 months	Spring-Autumn
Duck	300-350	6-7 months	Spring-Late Summer
Goose	20-40	9 months to 2 years	Spring
Guinea Fowl	80-160	8 months	Spring-Autumn
Pheasant	40-50	6 months	Spring-Summer
Quail	300	5-8 weeks	All year, unless moulting
Turkey	100	7 months	Spring

Chicken

Most research available has been in relation to chickens, as they are the most widely kept fowl across the world. Commercial producers and governments, via organisations such as the Commonwealth Scientific and Industrial Research Organisation (CSIRO) in Australia, the Department for Environment, Food and Rural Affairs (DEFRA) in the UK, the United States Department of Agriculture: Agricultural Research Service (USDA: ARS) in America, and many universities across the world, have undertaken extensive research into chickens. I have included findings relevant to chicken eggs in this section.

Size

Many factors affect the size of an egg.

- Egg size varies in size for both bantam and standard breeds.
- True bantam breeds (e.g. Sebright, Rosecomb) average a 29g egg.
- Pullet eggs are often smaller with stronger shells.
- For birds of the same breed and size, the larger the egg, the fewer eggs are laid.
- As a general principle, egg size is proportional to body size. Smaller hens lay smaller eggs and vice versa for large hens.
- In a clutch of eggs laid by the same hen, the first egg is often the biggest. Identifying hens that lay a consistent size is of benefit when selecting matching eggs for the exhibition plate.
- Egg size is affected by the thyroid gland, and in hot weather, the increased activity of the thyroid results in smaller eggs as the hens are prone to laying more often.
- Lighter breeds, which are more active, tend to have smaller eggs, and this may be due to higher thyroid activity compared to the heavier, more docile breeds.
- Disease can reduce egg size. Therefore, ensuring the health of hens is of prime importance.

Colour

There is significant variation in colour depending on the breed of chicken. Cross-breeding can be done to achieve different colour combinations, such as Araucana with a dark laying breed like the Maran, resulting in olive-coloured eggs.

White Eggs

The shell membrane is derived from the hemin (haemoglobin) of the blood and is responsible for the pinkish cast noticeable on the inside of freshly opened white eggs.

Brown Eggs

- The genetic basis for brown is affected by multiple factors; some are sex-linked. Daughters of a white-shelled male lay lighter eggs than those of the male with brown shell genetics.
- Brown colour is only on the outer coating of the shell (cuticula) and can be easily marked, so careful handling is essential.
- Shell colour changes markedly with the season, darkest when birds come into lay, and gets paler as the laying season progresses.
- Shells are darker in the first 10 eggs laid after a moult than in the last 10 laid before.
- Planning when your birds come into lay to coincide with a show could be of benefit when wanting to get the darkest brown eggs.
- The personal experience of the author breeding and showing Welsummer brown eggs for over 20 years shows that more consistent and darker eggs have been achieved by ensuring green feed is always available.

Blue or Green Eggs

- Gene O is the gene for blue egg colour and is dominant, so if the bird only has a single O gene, it will lay a blue-shelled egg.
- The effect of gene O goes entirely through the shell of the egg, as the colour is derived from the green/blue of the bile duct and binds with the shell as it is created.
- Variation of colour occurs between birds of the same age and breed. Running egg trials to identify the best colour layers for showing is worth the effort for future breeding and to ensure good quality-coloured eggs.

Olive Eggs

- Crossing dark laying breeds with blue egg laying breeds results in the offspring laying olive-coloured eggs, as the blue shell has a brown coating. The father should be from the dark layers and the mother a blue layer. Examples are Welsummer/Araucana cross.
- The depth of colour is primarily related to the amount and type of porphyrins (pigments) deposited on the eggshell.

Shape

- The shape of the egg is hereditary, making the selection of hens that lay the best shape important. For example, Silkies tend to lay rounder eggs.
- Pullets lay a longer egg initially and then a broader egg over time, due to the stretching of the isthmus region of the oviduct.
- The reason for the ideal shape in the standard is to encourage breeding and selecting for the viability of an egg hatching.

Shell

- Shell quality is affected by hereditary traits. Selecting for consistent quality of eggshell not only provides a better egg for the exhibition plate but also greater viability when hatching.
- The shape of the egg affects the thickness.
    - Round eggs have a thicker shell.
    - Longer eggs have a thinner shell.
- Shells become thinner in hot weather.
- The darker the brown egg, the thicker the shell, most likely due to the slower rate at which the egg is formed, not only producing a thicker coating of cuticula, but also a thicker shell.

Albumen

- The proportion of firm albumen is inherited
- Riboflavin is present in albumen and is responsible for the greenish-yellow tinge. It is an essential nutrient for chick development. Ensuring good feed is provided will improve the quality of the egg contents.
- "Fishy-smelling" eggs are not caused by feeding but a malfunction of the hen. Identify the bird and look for disease.
- There is no evidence that small blood spots or meat spots are hereditary. Anecdotally, the author has found that brown eggs are more prone to meat spots because the detection of them by candling can be difficult. The meat spots of brown eggs are darker and therefore more noticeable than those from a white-shelled egg.

Yolk

- The yolk is proportionately larger in small eggs compared to large eggs.

- The size of the yolk varies due to factors such as the breed and age of the hen, along with nutrition.
- Uniform shade of colour with no mottling is the ideal. The colour of the yolk is determined by the carotenoids in the diet. These come from the pigments of plants. Feeding a consistent diet will help to reduce mottling.
- Yolk colour can change as the egg ages.
- Double yolk and multiple yolks can and do occur, particularly for young pullets coming into lay or older hens that have reached menopause. Noting which birds are laying a particular egg will help when selecting eggs to show. Generally, multiple-yolked eggs are considerably larger than the average laid for that hen and are rounded at both ends. They often occur with a definite, depressed ring around the middle of the shell. The single egg is ideal for both hatching and for recipe consistency. Therefore, multiple-yolked entries will be disqualified.

Laying

Be aware of when a hen is about to lay. Behaviour will include being more restless, becoming busy arranging a nest, or lying quietly in a quiet, dark corner of the pen or nest box; often squatting when approached and being more vocal. Avoid disturbing birds that show signs they are about to lay, as you may cause them to delay laying and potentially spoil the egg.

For chickens, the entire time from ovulation to laying is about 25 hours. Around 30 minutes after laying, the hen will ovulate and begin the process again. As a result, they lay about an hour later for each successive day. This is useful to know when trying to collect eggs, before they have the chance to spoil in the nest.

Chicken Egg Colour by Breed

Colour	Example	Breed
Dark Brown		Bamevelder Langshan Maran Welsummer
Light Brown Tinted		Brahma, Naked Neck New Hampshire, Plymouth Rock, Orpington, Rhode Island Red, Sussec, Wyandotte, Yokohama
Blue Green		Araucana Cream Legbar
Olive		Cross-breed a dark brown layer with an Araucana Hen
Off White Cream		Asil, Cochin, Dorking Faverolles, Ixworth Leghorn, Silkie
Pink Plum		Croad Langshan
White		Ancona, Andalusian, Campine, Crevecoeur, Dutch Bantam, Lakenvelder, Minorca, Spanish Sicilian Buttercup

This is **NOT** an exhaustive list. Variation within a breed and cross-breeding will affect shell and cuticula colour.

Breed Specific Egg Standards

There are not currently any egg standards specific to a particular breed in the Australian Standard at the time of publication.

There is, however, a standard for Welsummer eggs in the British Standard. Contact the breed club of the birds you are interested in exhibiting eggs and ask the following:

- Do they have an egg standard?
- Is there a shell colour chart?
- Do they know of shows holding breed-specific egg classes?

These are examples of breed clubs around the world that have a colour chart and or an egg standard.

- Araucana
- Maran
- Welsummer

Some breed clubs publish a printed colour scale by which to measure the depth of colour. Any variation from the specific standard may be penalised.

USA Maran Colour Chart

UK Maran Colour Chart

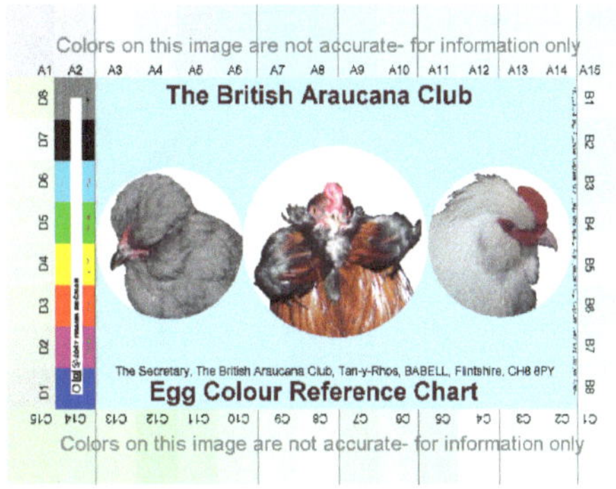

UK Araucana Colour Chart

Welsummer Egg Standard

The UK and Australian Welsummer Clubs have given permission to publish this standard and colour chart to encourage the showing of Welsummer eggs, known for their large size and dark brown colour.

Scale of points	
Colour	25
Shape	25
Size	20
Shell Texture	20
Appearance and Bloom	10
Total	100

Colour: A rich, deep red-brown, as dark as possible. The pigment is to be evenly distributed over the whole surface. Some products are speckled, mottled and occasionally blotched.

Shape: Egg-shaped; the top, containing the air space, domed, the bottom less so and more pointed, with ample girth.

Size: Exhibition eggs should be of a good size. Eggs should follow the weight standards laid out in the egg section of the standards.

Shell texture: Matt, smooth, and free from ridges, pimples, or porosity. Glossy eggs can be produced, but the matt egg is preferred.

Appearance and bloom: Exhibition eggs should be fresh, clean, with new-laid bloom and with minimal nest marks and scratches.

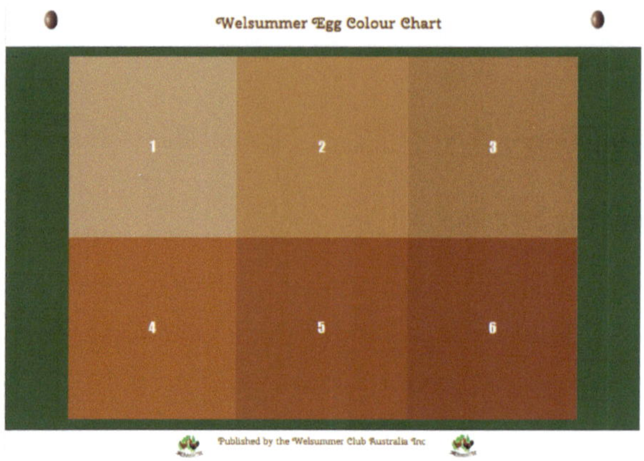

Serious Defects

- Pale colour
- Poor shape: spherical, narrow, or equally domed at both ends.
- Small size
- Uneven shell texture: ridges, calcareous pimples, or roughness at either end.
- Very glossy or thin and porous shell (**Note:** Silkies have glossy eggs).
- Excessive nest marks or scratches.
- Dirty or stained
- Anything interfering with hatchability.
- Staleness
- When more than one egg forms a single exhibit, they should match and be similar in all respects. Failure to do so constitutes a serious defect.

Duck

Duck eggs are distinctly different from land eggs. The shell, membranes and albumen are designed to protect the duckling in a wet environment where the existence and transmission of bacteria is far more prevalent than on dry land.

Shape

Duck eggs have a variety of shapes depending on the breed.

Most ducks breeds lay eggs longer than chickens.

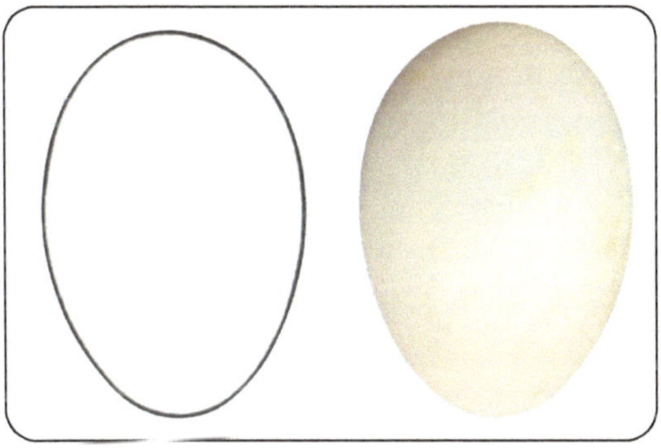

Comparative Size

A chicken egg is two-thirds the size of a duck egg.

Shell

The shell and internal membranes are much thicker compared to chicken eggs, making duck eggs much stronger. Therefore, it is much harder to crack duck eggs compared to chicken eggs.

The shell of this duck egg shows lines of stress on the shell, which are easier to see when candling the egg.

Duck Breed Variations

- Mallard duck eggs are proportionately longer and thinner.
- Muscovy eggs also have a wax cuticle providing another layer of protection. This wax coating can be removed as preparation for competition.
- Muscovy is the only breed of duck that generally goes broody.
- There is a greater variation in incubation time for ducks, depending on the breed.

Albumen and Yolk

The albumen is much thicker, so the yolk is better protected, which also means duck eggs often have better yolk eccentricity (central position in the albumen).

Waterfowl albumen must be clear. It is therefore best for duck eggs to be entered into the waterfowl classes.

In this picture, you can see that the yolk takes up a much larger proportion of the contents of the egg.

Time of Lay

Ducks lay between 4 a.m. and 7 a.m., so it is best to collect eggs first thing in the morning before they become soiled in the nest.

Shell Colour

Duck eggs have a more even application of colour compared to chicken eggs. The colour of eggshells varies considerably for an individual bird, and you may see more variation across the same breed.

The table shown is not an exhaustive list, just an indicator of what colour types are available for duck eggs.

Duck Shell Colours

Colour	Example	Breed
Black		Cayuga
Dark Grey		Black East Indian
Light grey		Watervale
Light Brown		Orpington
Light Green		Rouen Bali
Blue		Australian Call Duck Sweidsh Blue
White		Aylesbury

It is important to run egg trials and check the colour when selecting birds, as the bird may not lay true to breed.

Cheryl Rowles: Cayuga Duck Eggs

Freshly washed Cayuga eggs, showing that the first laid of the season is the darkest in colour.

Breeds of Duck in Australia

Abacot Ranger
Australian Call
Aylesbury
Bali
Black East Indian
Blue Swedish
Campbell
Cayuga
Crested
Elizabeth
Indian Runner
Magpie
Mallard
Muscovy
Orpington
Pekin
Pommern
Rouen
Rouen Clair
Saxony
Silver Appleyard
Watervale
Welsh Harlequin

Goose

Goose eggs have thicker shells, a higher yolk-to-white ratio, and a richer, fattier yolk with a stronger flavour than chicken eggs, making them more nutritious and flavourful.

Goose eggs contain higher levels of protein, vitamins A and E, and certain minerals compared to chicken eggs, according to nutritional analyses.

Goose eggs from Roman geese or Chinese geese are considered among the best for eating due to their flavour and texture.
The harder shell makes them much harder to crack than duck or chicken eggs.

Shape

Comparative Size

A goose egg weighs between 143 and 185g and is the equivalent of two or three standard chicken eggs.

Shell Colour

All breeds of geese lay WHITE.

Geese have a very short laying season. They start to lay two weeks after the winter solstice. They produce 20-40 eggs in a season, depending on the breed of goose and the age of the bird.

The larger egg and stronger shell make goose eggs ideal for painting or ornamenting.

Egg Proportions		
Egg component	**Chicken**	**Goose**
Whole Egg	57g	218g
Egg Shell	8g	50g
Egg Yolk	17g (34%)	87g (52%)
Egg White	32g (66%)	81g (48%)

The difference in contents needs to be accounted for, which is why waterfowl are judged in their own class.

Breeds of Geese in Australia

African
Australian Settler
Brecon Buff
Chinese
Embden
Pomeranian
Tufted Roman
Sebastopol
Toulouse

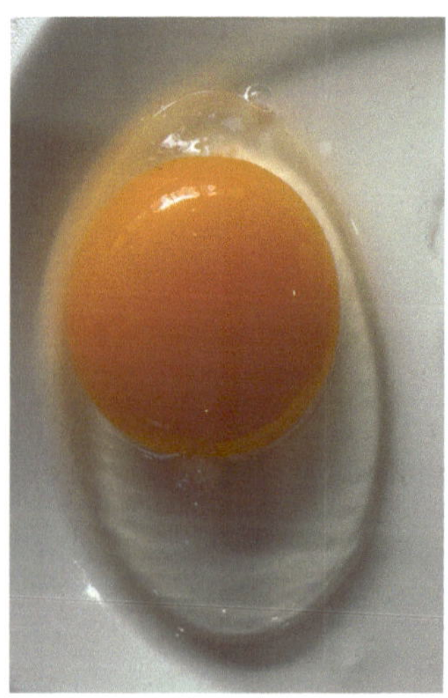

Guinea Fowl

Compared to a chicken egg, the shell of the guinea fowl is very thick and, therefore, more difficult to crack. The eggs contain a higher proportion of yolk to white compared to chicken eggs.

Shape

Guinea fowl eggs are smaller and pointier on one end than a normal chicken egg, making them conical in shape.

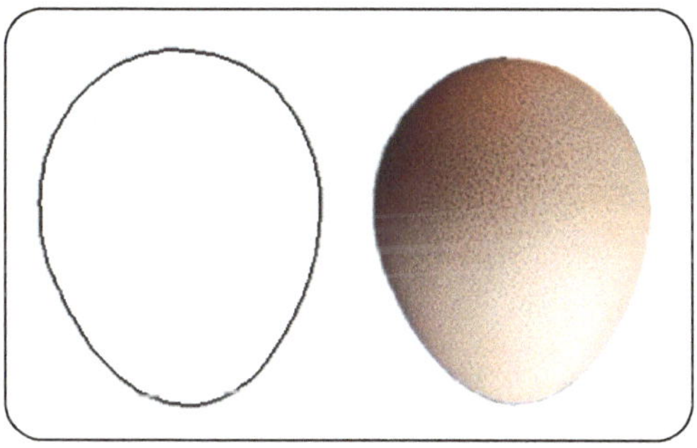

Comparative Size

It is two-thirds the size of a chicken egg, around 40g on average.

Shell Colour

Cream to light brown with faint to dark brown speckles.

Guinea Fowl are seasonal layers, laying daily between spring and autumn. If you want to ensure you have a supply of eggs, you will need to keep them in a coop and collect daily.

Guinea Fowl lay on the ground. Keeping the floor covered with wood shavings or sand will help to keep the eggs clean.

Free-range hens will lay in a communal nest amongst the shrubs and long grass. You will need to hunt the eggs out regularly if you want some to exhibit. They lay between 9 a.m. and 11 a.m., so this is a good time to find out where the nest is.

Varieties of Guinea Fowl in Australia

Only the Helmeted Guinea Fowl can be found in Australia with the following colours:

Cinnamon
Lavender
Pearl
Pied
White

Pheasant

Only the Jungle Fowl amongst the pheasant species are listed in the Australian Poultry Standards. Pheasants do not respond well to being in small cages or being handled by a stranger, such as a judge at a show. However, you can enter their eggs.

Shell Colour

Varies by breed, most commonly olive-brown, but can have brownish or blueish tones

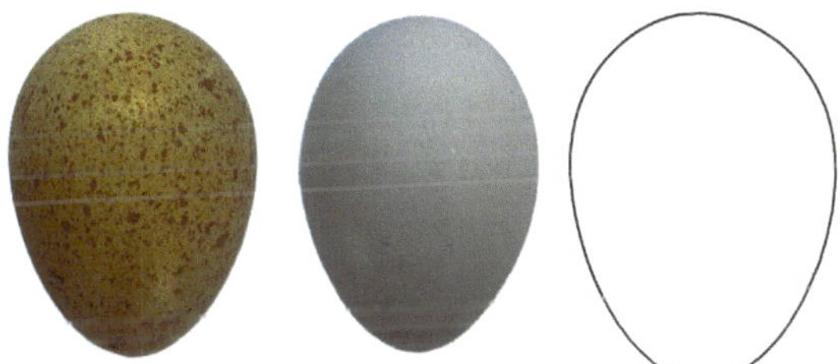

Chuck Partridge Lady Amherst Pheasant

The number of eggs in a clutch varies across the breeds of pheasant. The age of the bird and the strain (genetic lineage) will affect the number of eggs in a laying season, getting as low as 20 or as high as 130 per hen. Average egg production is about 50 per hen per season. Egg production is slightly higher in second-year hens when compared with first-year hens, but egg numbers decline rapidly after this.

Laying starts in spring and goes into summer. This limits the time of year you can have eggs to exhibit, unless you set up lights to prolong the laying season.

Thanks to Doug Sommerville for providing information on pheasant eggs.

Breed	Egg Laying
Chuck Partridge	Prolific layers
Peahen	4-6 per clutch
Ring Neck Pheasant	50-70 per season
All other pheasant breeds	4-20 in a clutch

Freshness

Determining freshness externally will be by the size of the air sac when candled.

Comparative Size

The same as a bantam chicken egg, less than 42.5g. Except for Peafowl, which are a third larger than chicken eggs.

Breed	Length (mm)	Breadth (mm)	Weight (g)	Shape Index
Chuck Partridge	37.3	27.5	20.8	73%
Lady Amherst Pheasant	45.4	32.3	32.8	71%
Peafowl	76	56	105	73%

Breeds of Pheasant in Australia

Chuck Partridge
Golden
Jungle fowl
Lady Amherst
Nepal Kalij
Peafowl
Reeves's
Ring-Necked (Game)
Swinhoe's Pheasant

Golden Pheasant

Quail

The nutritional value of quail eggs is similar to that of chicken eggs, as quail eggs are so small, you eat more of them to get the same nutrition.

Shape

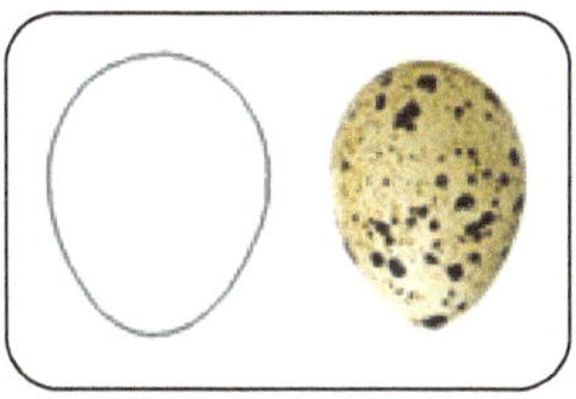

Comparative Size

About three to four quail eggs are the equivalent of a single chicken egg.

Shell

Quail egg has a texture where the spots are prominent; this is considered normal.

Shell Colour

Most commonly speckled eggs with large blotches; however, some lay plain eggs without speckles.

Quail eggs taste very similar to chicken eggs. However, they have a more significant yolk-to-white ratio and are creamier and taste richer than chicken eggs.

Storage

Storage time for washed eggs is less than storage time for unwashed eggs. This is because the protective bloom has been removed.

Quail eggs stored in the fridge:

- Unwashed eggs can be stored for three months.
- Washed eggs can be stored for six weeks.

Note: Quail eggs are very fragile compared to chicken, so take great care when collecting. Quail lay later in the day between 2-6pm

Breeds of Quail in Australia

California
Japanese
Quartix

Turkey

Turkeys will lay their eggs anywhere, so you may need to hunt high and low, including the roofs of pens for eggs! Turkey shell membrane forms a greater percentage of the egg than chicken eggs, making these eggs tougher to crack.

Shape

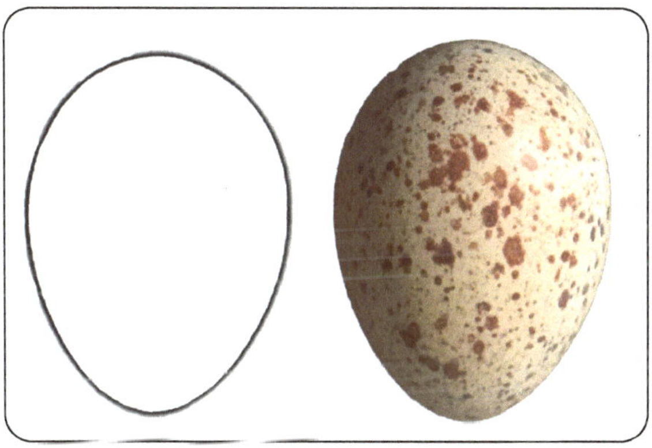

Freshness

Freshness is more difficult to assess as there is no obvious bloom, nor a colour that fades. Therefore, candling or the float test can be used without cracking the egg to check the air sac.

The stronger shell membrane enables turkey eggs to stay fresher with a longer incubation period of 24 days compared to chickens' 21 days.

Comparative Size

Turkey eggs average 70g to 100g and are bigger than duck eggs and smaller than goose eggs.

Composition: Yolk 30.53% and Albumen 54.17%

Shell Colour

Turkey eggs can vary from light to medium dark brown and are sprinkled with medium to dark brown spots. However, it has been known for Turkeys to lay eggs without speckles.

A strong shell free from blemishes speckled that are reddish/brown, with a white to deep cream ground colour, sometimes with a pink tinge. Evenness of ground colour is important. The more uniformity of colour and density of speckling, the better.

NOTE: Blotches or mottling of colour is considered a fault.

Fertile Eggs

Determining if an egg is fertile is more difficult with turkey eggs, as the halo is not so clearly defined as it is in chicken eggs. It is a chalky white spot.

Classes

If there are no turkey classes in the schedule, then the eggs should be entered into the class with the same ground colour.

Varieties of Turkey in Australia

There is just one breed of turkey; however, they come in various colours.

Black
Narragansett
Blue
Bourbon Red
Bronze
Buff
Royal Palm
Slate
White

Photo: Melissa Hall. Ambleside Photography

Prepared Egg Classes

The decoration of eggs has occurred all around the world for millennia, with the oldest egg engraved with patterns dating back 60,000 years, found in a cave in South Africa.

In more recent years, painted, decorated, and displayed eggs have grown in popularity at poultry and agricultural shows. Prepared eggs are not in the Australian Standard as of December 2025. In the UK, a scale of points out of 100 has been developed to cater for these entries.

Painted Eggs

The egg can be painted with any media — oils, watercolour, felt tip, or ink. There must be no other form of adornment. They are best displayed on rings 1.5cm deep, so the complete painting can be displayed. The top of a milk carton makes a good ring on which to display chicken egg.

Decorated Eggs

The blown egg can have a cut shell and is often painted, with the addition of decorations such as beads, stickers, sequins, and dried flowers. The shape of the egg **must** be recognisable.

Check the standard for the scale of points and the criteria on which to judge. Keep a copy for reference.

Instead of a plain ring stand for painted classes, you can make your own stand. It's particularly good when it matches the overall design.

Displayed Eggs

Usually, a group of six eggs is decorated with a theme. It can consist of identical eggs or eggs from different species.

Displayed eggs are graded in a group. Therefore, the ability to see the quality, size, and shape of the eggs is essential. It is important that eggs should not be hidden by the display. Check the maximum dimensions of the display allowed in the schedule.

Check the standard for the scale of points and the criteria on which to judge. Keep a copy on hand for reference.

Choosing eggs to decorate

Chicken eggs come in many colours, as do duck eggs. Pekin ducks are the best duck eggs to use for decorating.

Goose eggs are white and much larger than chicken eggs. They provide a larger surface on which to paint. The shell is much stronger, making it ideal for decorative carving.

Emu eggs have a thick, dark green shell, which enables intricate carved designs.

You can hard-boil a prepared egg, but you will only be able to keep the egg for a short time before it starts to rot and smell.

A blown egg is, a hollowed-out eggshell, created by piercing small holes in the top and bottom of a raw egg and blowing out the yolk and white. It can be kept indefinitely, and should you accidentally drop or crack the egg, it will not smell. You also remove the risk of the eggshell discolouring and spoiling the decoration.

How to Blow an Egg

You will need:

- Eggs not more than two weeks old, as fresher eggs are easier to blow.
- Tape — masking or clear, whatever you have available.
- Newspaper to cover the table. It can get messy!
- Kitchen roll or tissues.
- Egg cups.
- A container to collect the contents of the egg.
- Needle — an old sewing machine needle or a large, sharp sewing needle.
- Metal skewer or small knitting needle. Something round with a pointed end that is strong enough to pierce the shell.
- Sanitiser or bleach.
- Manual football pump or small air compressor, if you have one.
- Syringe, if you have one, but you can make do without.

Method:

1. Leave eggs at room temperature before blowing.
2. Wash the outside of the egg to remove any dirt or debris. Be careful with turkey eggs and those that are speckled. You need to leave the natural surface undamaged if you are not decorating them.

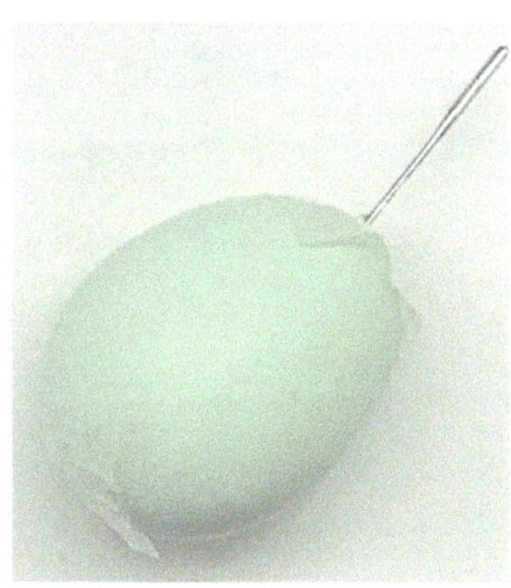

3. Place a square of tape on each end of the egg; this helps to stop the shell from cracking.
4. Pierce the top of the egg (pointed end) with one hole using the small needle; you want this hole to remain small.

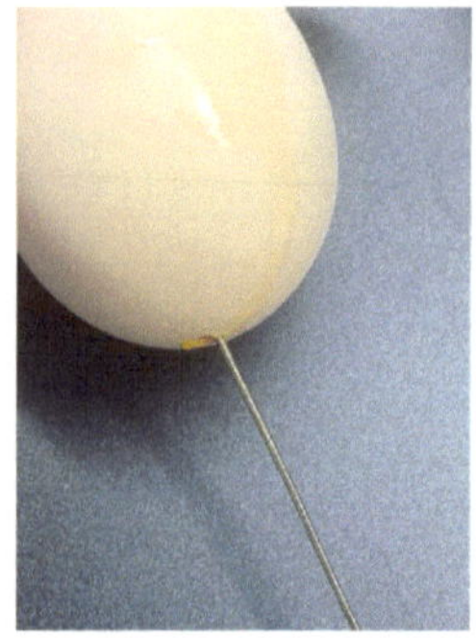

5. Use the small needle to pierce a hole on the other end of the egg to make the initial hole. Now, using a larger needle, make the hole bigger. Use the skewer/knitting needle to make an even bigger hole.
6. Poke far enough to break the yolk. Following this, gently shake the egg to mix up the contents.
7. Remove the tape from the pointed end.
8. Manual method — Over the container, place your mouth over the small hole, making a seal. Blow gently initially and gradually add more pressure. The contents will steadily start to come out at the bottom. You may need to make the bottom hole larger. Alternatively, use a hand pump or air compressor.
9. If you need to take a break from blowing, place the egg in the egg cup so it can continue to drain the contents.
10. Keep blowing until the egg feels very light and you think it is empty. Use the tissue to wipe the bottom so you can see if more yolk comes out.

11. Once the egg is empty, place it in a bowl of clean water, and use the syringe to fill up the egg with water. Give the egg a gentle shake. If you don't have a syringe, hold the egg under the surface of the water with a large hole to the top to allow air out and water in.
12. Take the egg out of the water to drain. You may need to blow the

egg again to get the stickier yolk out. Repeat until no more yolk comes out. It is important to remove all the contents, or the egg will smell, and the contents may stain the shell. Wash out with bleach or sanitiser to ensure the egg does not go mouldy.

13. Place the eggs pointed end up, with a large hole below to allow them to empty whilst air drying.
14. Once fully dry, you can decorate your egg.

Competition

Clubs and Competition

One of the functions of a poultry club is to organise competitions (shows). These may be annual poultry competitions or part of an agricultural show where other domesticated animals are also exhibited.

Chief Steward

The chief steward organises the poultry show by creating a schedule in which all the classes are listed. The approved schedule is distributed to exhibitors. The chief steward processes all the entries and allocates the pens for birds grouped by class. It is also important to ensure there are sufficient plates and tables for egg classes.

Schedule

- Classes should be listed clearly by species, size, colour, and number of eggs per plate.
- Egg numbers can vary by class. Allow for the size of the plates and the size of the eggs when choosing the number of eggs for a particular class. Only three duck eggs will fit on the average plate as compared to six large chicken eggs.
- White/cream (off-white)/tinted eggs could be in separate classes, rather than a "white only" class.
- Clear descriptions are given for colour definitions, as there is often confusion with reference to tinted eggs.

- Here are a couple of example definitions I found in the same show schedule:
    1. Tinted eggs are to be very pale brown to cream, including light beige, ivory, or off-white. Tinted does not include white, medium, dark brown, blue, or green.
    2. Tinted eggs are any other natural colour shell (pink, pink-brown, green, cream, speckled, etc.).

The classes in that particular schedule had both tinted and any other colour (AOC) egg classes.

When I contacted the chief steward, they confirmed that tinted meant pale brown and cream only. That any other colour was pink, pink-brown, green, non-white, and blue. So, if you have any doubts as to what class to enter your eggs in a schedule, contact the chief steward for clarification.

Common Entry Terms and Conditions

Standard Class Terms

- Eggs will be judged according to the Australian Egg Standard.
- Bantam eggs are not to exceed 42.5g.
- Egg stewards and egg judges should be named in the schedule.
- Only fresh eggs may be entered for standard classes.
- The maximum number of entries per exhibitor will depend on the overall capacity of tables and plates for egg classes.
- Mottled/speckled eggs will be judged by ground colour.
- The use of external and/or internal judging should be stated.
- Reclassification of eggs prior to judging will only be undertaken by the egg steward.

At the end of the show, confirm if the eggs will be returned to exhibitors or if they will be disposed of by the egg steward.

Non-Standard Class Terms

- It is customary that non-standard egg classes, such as display eggs or species not in the standard, can NOT be awarded best egg in show.
- A statement of the maximum dimensions of the displayed egg exhibits should be included. i.e., 30x30cm.
- Displayed eggs should not hide the quality and matching of eggs.

Standard Classes

Classes exist to ensure eggs of the same type are compared to each other for easier inspection and judging. Only the best in each class can be considered for "Best Egg in Show".

Classes are defined by size, species, and shell colour.

Size	Species	Shell Colour
Bantam Large	Chicken Duck Turkey Goose	White Brown Green Any other colour
	*Guinea Fowl *Pheasant *Quail	

* Although not mentioned in the standard, these eggs can be exhibited, unless specifically excluded in the show schedule. These non-standard classes are not considered to be eligible for best egg in show.

A standard for colour shades is included. This is not a limited list. All colours could be entered.
Example eggshell colours are shown in the shell colour section of this book.

Class Criteria

Size	Colour	Number of eggs per plate
Bantam	Dark Brown Light Brown Off-White/Cream Any Other Colour (AOC) Distinct Contrasting Colours (DCC)	1 3 6 12
Large	Dark Brown (including Maran and Welsummer) Light Brown Off-White/Cream Any Other Colour (AOC)	1 3 6 12
Mixed	MUST describe clearly in the show schedule, indicating exactly the specifications of the class. i.e., Mixed size Mixed species Mixed shell colour	3 6 12
Waterfowl and Turkey	*Duck *Goose *Turkey	#1 #3

Due to large egg size and lower frequency of laying, for these species, fewer eggs are required per plate.

* External judging only is most common for these species, as it is often impracticable to internally judge such a large egg.

Example Egg Classes

Chicken — six eggs per plate

Bantam

10	White
11	Cream/Off-White
12	Light Brown
13	Dark Brown
14	Any Other Colour

When creating the schedule, choose classes that give maximum opportunity for entries. Club members' breeds will attract more entries

Large

15	White
16	Cream/Off-White
17	Light Brown
18	Dark Brown
19	Any Other Colour

Mixed

20	Mixed, distinct, contrasting colours, same size
21	Any size, any colour, same species

Waterfowl and Turkey — three eggs per plate

22	Duck	All the same colour
23	Duck	Mixed plate, any colour
24	Goose	
25	Turkey	

Quail — 12 eggs per plate

26	Any colour

Clubs could have a separate egg show if the annual show is the wrong time of year for eggs.

Any Other Species/Field and Forest — one to three eggs per plate

27	Pheasant, Ostrich, Emu

Commercial Eggs

A commercial egg is an egg produced from a commercial breed of chicken. For example, the Hy-Line Brown and Lohmann Brown. These breeds have been developed to lay an egg a day from six months to 18 months of age.

Classes for commercial eggs are judged on a larger number of eggs in an entry. Normally, 18 eggs are divided into three plates of six. Each plate is judged separately; an average is calculated across the three plates to give the overall score for the entry.

Allowance needs to be made for the shape and size of commercial eggs. These usually have a more rounded shape than the ideal used in the poultry standard.

Consistent weight of eggs in a class is of particular importance as consistent quality is a criterion important to commercial egg farmers.

Specialised Breeders Australia is a company that raises and sells commercial chickens for both egg laying and meat production. They have kindly allowed me to include a picture of their promotional shell colour card.

Commercial egg farmers also select for eggs that **DO NOT** have spots or speckles on the shell.

School Competitions

School Egg Competitions started at the Sydney Royal Easter Show around 2012. They are only possible due to sponsorship by the Australian commercial egg industry body, Australian Eggs, in conjunction with Specialised Breeders Australia.

The primary or secondary school needs to have the facilities to keep poultry in order to enter the competition. All participating schools are supplied at the same time with Hy-Line pullets between five and eight weeks, along with an information pack and teaching guide.

The steward ensures that the birds are supplied with sufficient time, so they come into full lay at the show at around 22 weeks of age. The birds from each school are penned on the first day of the show. The eggs laid for each school are collected and held until judging day.

Judging is based on three elements. The school with the highest score across all three elements is deemed the overall winner.

- Project
- Hen presentation
- Egg

Project Class

During the time of caring for the birds, students also produce a written project on the keeping of laying hens, containing information on the egg industry. Primary and secondary school entries are assessed separately.

Hy-Line Class

The hens are judged according to commercial assessment criteria, checking weight for age as well as the general condition of the birds.

Egg Class

The eggs laid by the hens whilst penned at the show are collected for each school and judged.

Preparation

Egg Steward

The egg steward is responsible for running the egg section of a show. Egg stewards take responsibility for the list of entries, ensuring names remain confidential. The judge should have no knowledge of who entered the eggs.

The steward sets up all the equipment, such as tables and class markers; accepts the egg entries from the exhibitors; and ensures that all eggs are plated up into the correct class.

During the judging process, the steward marks off the position achieved for each entry in a class, continuing to ensure anonymity of entries to maintain a fair and equitable judging process.

Before the Show Checklist

- **Tables** — Check there are sufficient tables and plates to cater for the maximum number of entries (30cm is the ideal, check the plate size). Allow additional plates for internal judging if indicated on the schedule. Tables should not be too deep, or judges will struggle to reach plates at the back of the table. Alternatively, provide access to both sides of a deep table.

- **Plates** —
    - **External Judging** — Ceramic plates, plastic garden pot saucers, and disposable plates can be used. It is important that they are plain with no pattern.
    - **Internal Judging** — Separate white ceramic or plastic plates are ideal as they allow the egg contents to flow and lay flat. Paper plates can be used; however, they are porous, and the contents dry out quickly.
- **Sawdust/Dried Sand/Grain** — Use on the external plate so the eggs can be positioned and not roll around the plate.
- **Hand Wipes or Tissues** — Allows the judge to clean their hands after cracking eggs for internal judging.
- **Tablecloths** — Plain dark green is ideal; white paper will suffice. Regardless of colour, providing a plain-coloured clean surface makes judging easier.
- **Class markers** — Tape or string to indicate the start and end of each class.
- **Cards** — Consider printing smaller cards for eggs to reduce overcrowding.
- **Marking Sheet of Entries for the Class Steward** — Don't forget that the "Egg of Show" and reserve awards need to be included for standard classes ONLY.

On the Show Day Checklist

- **Tables** — ensure the tables are level and there is good lighting. If possible, position plates out of the reach of the public, especially prior to and during judging.
- **Exhibits** — should only be handled by the judge or steward after entry.

- Each entry should be staged on identical plates (preferably plain white with no decoration). A little fine sawdust/dried sand or grain can be used to arrange the eggs to ensure they do not roll around the plate.
- Do not overcrowd the table, as room is needed for judging.
- Each plate should be clearly marked with the entry number.
- DO NOT fix plates to the table; this will allow the judge to move exhibits easily for close examination.
- Have either wipes or clean water and soap to allow the judge to wash their hands. This is important for internal judging as their hands can get sticky with albumen.
- Judges should not be allowed to reclassify eggs.
- Judges should NOT be present whilst eggs are collected from exhibitors and plated up.
- Only the steward should see the names of the exhibitors for specific entries to ensure all classes are judged without bias.

Moving Classes

Only the steward has the authority to move eggs to a different class.

If, for example, there is NO turkey class, then turkey eggs could be moved to a class based on the base colour of cream/tinted.

Judge

Megg Miller

Judging

The role of the judge is that of an independent appraiser. Judges can be chosen from a panel of registered judges. Each country/state will have its own governing body and process for becoming a judge.

In Australia, each state poultry body keeps a list of qualified and provisional judges. In the UK, the Poultry Club is the central organisation that oversees registration of judges. In the USA the American Poultry Association (APA) is the organisation that oversees judge registration.

The judge will use their experience and training to appraise the eggs both externally and internally according to the poultry standards and the rules printed in the schedule for that show.

Standards

The information here is gathered from a variety of sources and is based on the author's experience of judging eggs. No one method is preferred. Different methods have been described to provide options for measuring and judging the quality of an egg.

I do, however, encourage you to read and understand the standard for eggs in the country in which you are judging. Keeping a copy of that standard at hand whilst you judge as a prompt will help you appraise consistently across all exhibits.

Judging Aids

This is NOT an exhaustive list; this is merely a guide of the items to assist with judging.

- **Egg scales that measure less than a gram**
  (Weight carries the most points in the scale of points, so determining the weight of the eggs is essential to judging.)
- Box/bag to keep all your judging aids together
- Copy of the current poultry standards
- Colour fan or chart of yolk colour
- Colour chart of shell colour
- Score cards
- Notepad
- Pen/pencil
- Egg anvil or broad knife
- Calculator
- Egg judging tray (with both a green and black back cloth)
- Magnifying glass
- OvaView or OvaScope egg candler or high-intensity torch

TIP: Check the schedule of the show for which you are judging. Confirm if both external and internal judging are stated in the schedule. If not indicated, check with the steward prior to the show.

Here is a list of aspects of the egg you should consider checking for when judging.

External

- Weight
- Uniformity of size and shape
- Shell texture
- Colour
- Freshness, bloom, and appearance
- Air space

Internal

- Freshness
- Yolk colour
- Two layers of albumen
- Chalazae
- No meat spots

Scoring Cards

Whilst a scoring card might not be mandatory, it is a useful aid to judging eggs consistently across not only the classes but also each section. You can develop your own version to suit your judging style and methodology.

The table of nomenclature at the beginning of the book has a judging order column, which you may find useful for consistency in assessment.

Note: In the UK standards, an extra five points can be added for each egg for matching and uniformity. This option is NOT available in the Australian standards.

Example

Commercial exhibitors will often ask for feedback on their eggs; a scorecard can also serve as an aid when asked later for any feedback.

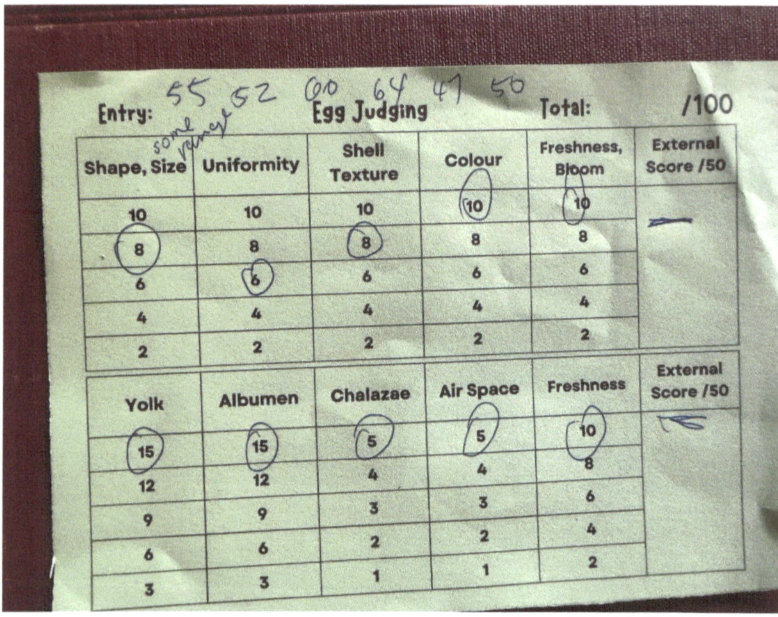

Entry: Egg Judging Total: /100

Shape, Size	Uniformity	Shell Texture	Colour	Freshness, Bloom	External Score /50
10	10	10	10	10	
8	8	8	8	8	
6	6	6	6	6	
4	4	4	4	4	
2	2	2	2	2	

Yolk	Albumen	Chalazae	Air Space	Freshness	External Score /50
15	15	5	5	10	
12	12	4	4	8	
9	9	3	3	6	
6	6	2	2	4	
3	3	1	1	2	

Thanks to Cathy Newton for sharing her egg scorecard

Breed-Specific Characteristics

Where a standard exists for a particular breed, such as Welsummer, Maran, Araucana, etc., compliance with that standard is required for any eggs in a class marked in the schedule specific to that breed.

Species-Specific Characteristics

The size of the yolk is affected by the age and the breed of the bird. Duck, turkey, and goose eggs have a much larger yolk by proportion compared to chicken eggs. This should be allowed when judging internal contents.

The thickness of albumen is much thicker for waterfowl and clearer.

Mixed Plates

A mixed plate can be for the same or different species/breeds with different colours. Check the schedule to ensure what has been defined for this class.

If it is unclear exactly what the intended criteria were for this class, check with the steward before commencing judging in order to assess with the correct criteria.

Example Class Definitions

1. A Mixed plate exhibit must be presented as a mix of different egg types, including, but not limited to: duck, fowl, goose, guinea fowl, pheasant, quail, turkey, etc.
2. A plate of eggs from the same breed with different shell colours, with at least two eggs of each colour.

Same Species — Different Colour Shell

- Look at the quality of each egg — shape, shell, and freshness.
- Matching and uniformity — the closer the uniformity, the higher the exhibit should be marked.

Mixed Species

- Look at the quality of each egg — shape, shell, and freshness.
- Take into consideration the characteristics of each species of egg.
- Matching and uniformity will not be possible due to the distinct differences between species.

Standard Classes

Eggs can be assessed both externally and internally, but should be judged separately, using points of scale from the standard.

External Judging

Weight

- Weight — use egg scales to check that the eggs are within the weights for the class.
- Attention should be paid to the bantam class, ensuring they are not overweight.
- Check for consistent weight of eggs on a plate — this helps to check matching and uniformity.

Shell

Cuticle/Bloom

Examine the cuticle closely, checking that the bloom is without dull or stale appearance.

- Lighting — use good lighting, preferably with a daylight bulb.
- Shell — pick up the egg and feel all over the surface for a smooth and even texture.
- For duck eggs, the air sac is often visible and should not be penalised when judging.

Shell Colour

- Check for artificial bleaching of the shell, especially white eggs. Use your sense of smell to detect chemical odours.
- Shell colour — If possible, use a dark green or black cloth as the background over which to compare eggs. Alternatively, use a white tablecloth.

Clare Beebe, a well-known UK egg judge, uses different back cloths on her Judging Board when looking at the shell colour.

> *"I prefer **black** for white/cream/olive/green/blue eggs and **green** for all shades of brown and plum"*

- Cream (can also be described as "off-white" or tinted), tones of shell colour should not be judged in classes for "white eggs".

Good, Even Colour   Poor Distribution of Colour

NOTE: As a judge, you cannot move eggs to a different class if you believe they are in the wrong colour class; you will need to confer with your steward and leave it to them to decide if they can be moved.

Shape

This is the ideal chicken egg. Check that the shape is even around the egg by rotating it in your hand around the sagittal/transverse axis.

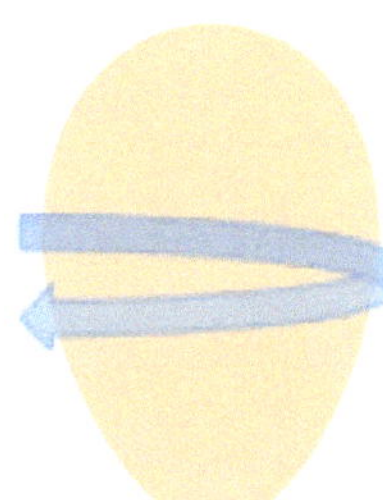

Know your Shapes

Be familiar with the shape of the egg for the species and what the ideal shape should be, not to scale.

Matching and Uniformity

There is more than one method to check for matching and uniformity for a plate of eggs.

1. Egg shape index
2. Line-up
3. Egg carton

Method 1 – Egg Shape Index

Callipers can be used to measure the length and width of an egg. This method is a more accurate way to measure uniformity, but it is also more time-consuming.

A scoring card is a helpful way to keep track of the measurements of each egg. You can work out how close to the ideal shape each egg is by taking the measurement of the breadth and length of the egg. The mathematical ideal for a chicken egg is when the width is 74% of the length.

To calculate the egg's actual ratio, use the following formula:

$$\frac{\text{Width (Horizontal plane) mm}}{\text{Length (Sagittal/Transverse plane) mm}} \times 100 = \underline{\qquad\qquad}\%$$

The egg shape index varies by species and breed. Here are some examples.

Shape Index	
Chicken	74%
Large Duck (Breed dependent)	70-75%
Bantam Duck	71%
Guinea Fowl	76%
Goose	70%
Pheasant (Breed dependent)	71-73%
Quail	78%
Turkey	78%

Method 2 – Line-Up

This is a quicker method to compare the size and shape of the eggs but is less accurate than the method using callipers. You will need a straight edge against which to line up the eggs.

In this example, a wooden ruler is used on a flat surface, with the blunt end of the egg set against the ruler. Alternatively, a straight wooden rod or dowel can be used.

Method 3 – Egg Carton

This method is used in commercial environments when assessing eggs laid.

Use an egg carton that has the low-profile separators for ease of viewing the egg. If you can't find a low-profile egg carton, then cut down the separators. Place the eggs in the carton pointed end down. Look at the side of the carton for comparison of the egg heights and the width of the egg.

In this picture, one egg stands taller than the others.

Freshness

Gentle Shake or Jerk Test

This method is used to try to determine if the albumen of the egg is firm, an indication of freshness.

Method

Hold the egg upright with the blunt end to the top, close to the ear. Jerk (quick, short movement with a sudden stop) the egg. Listen carefully for a sound like a flip. Feel if the contents wobble, as this will indicate that the albumen is soft and watery. This is an indication that the egg is not fresh.

Note: Don't overdo the jerk or repeat it too many times, or you can inadvertently make the egg "rattle"! A rattle occurs when the chalazae have broken from the yolk or the yolk sac has burst. If this occurs and an internal test is performed, then the egg will be considered to have a serious fault, and the entry will be disqualified.

Float Test

Placing eggs into a container of water and seeing which ones float is a known method, but it is impractical when judging large numbers of eggs.

Internal Judging

Break at least one egg onto a plate (NOT into an egg carton!) as it is important to see how the yolk sits within the albumen. This can only be achieved on a flat surface.

Methods to Crack Open an Egg

Method One

Use a stout, heavy knife to strike the shell.

Method Two

Strike the egg against an egg anvil. This could be made of wood or metal.

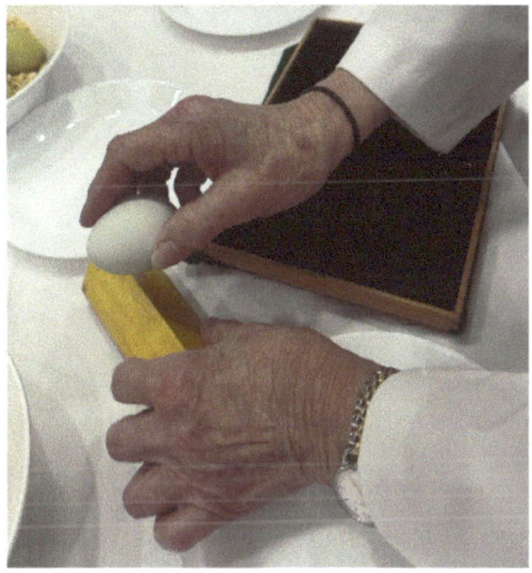

Crack the shell of the egg along the transverse plane to split the shell into two equal halves. (see diagram)

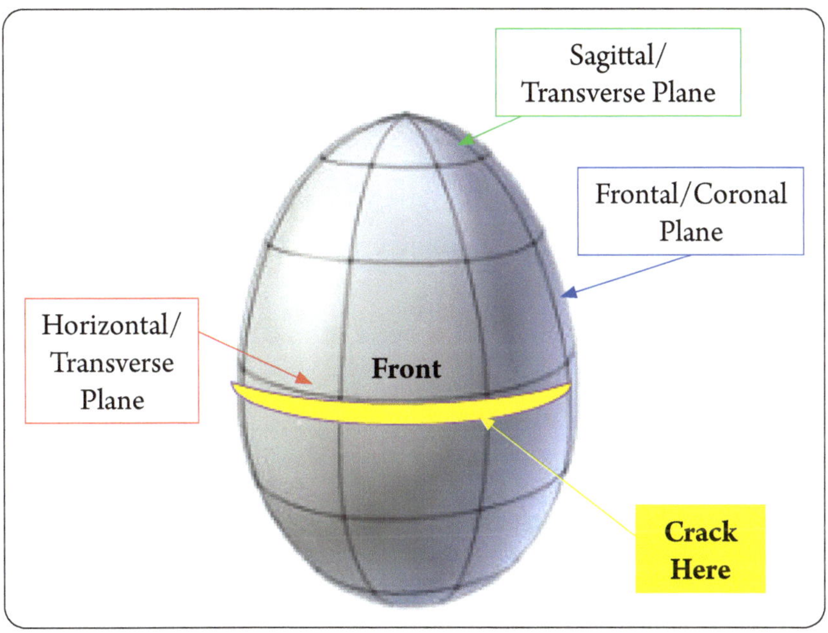

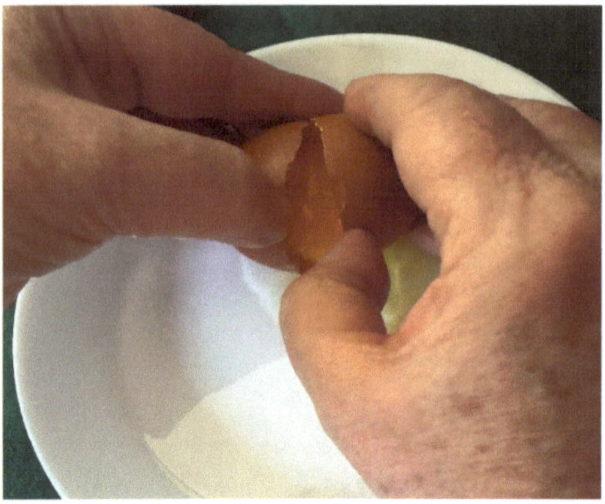

As the egg rests on the plate, gently pull apart the two halves of the shell to let the contents flow onto the plate. The intention is to avoid breaking the yolk.

Shell and Membrane

Air Sac

You can visually check for freshness by examining the air sac at the blunt end of the egg. If the air sac sits to one side and is small with a tight, unwrinkled surface, freshness is indicated.

You can use the callipers to measure the depth of the air sac. This measurement is used in commercial environments for chicken eggs, with the top graded eggs marked as special grade, where the distance is 4mm or less, and Grade 1 is equal to 8mm or less.

Albumen

Methods used to measure freshness are described here.

Measurement Techniques and Tools

There are two methods described,

1. Using a Haugh Micrometre
2. Using callipers

Method One – Haugh Unit

Haugh Micrometre

The height of the albumen can be measured for a more accurate comparison of albumen quality. In the commercial world, an instrument called a Haugh Micrometre is used to measure the height of the albumen and yolk.

The significance of the **Haugh unit** testing method takes into consideration the size of an egg (measured by its weight) so that the relative albumen height is calculated and compared.

The availability and cost of a Haugh Micrometre mean that very few judges own such an instrument. Should judges have access to a Haugh Micrometre, they are most likely to only use it for the contenders for the top awards of best egg of show and reserve egg of show, rather than to be used for every exhibit where the contents are judged. It is time-consuming to use.

The following methodology is recommended when using a Haugh Micrometre.

1. Weigh the egg in grams
2. Crack the egg onto a flat glass plate.
3. Position the albumen height gauge above the egg white, ensuring the measuring needle is approximately 1cm from the edge of the egg yolk.
4. Press down on the needle and note the unit of measure for the height.
5. Then run the calculation for the Haugh Unit.

Haugh Unit of Measure

The height, correlated with the weight, determines the **Haugh Unit**, or **HU**, rating. The higher the number, the better the quality of the egg (fresher).

The formula for the **Haugh Unit** is

$$HU = 100 \cdot \log_{10} (h - 1.7w^{0.37} + 7.6)$$

HU = Haugh unit (Range 0-130)

h = observed height of the albumen in millimetres

w = weight of egg in grams

Egg Commercial Grade	Haugh Unit
AA	72 or more
A	60-71
B	31-59
C	30 or less

Method Two – Albumen Index

Measuring the albumen index ratios is a standardised method of measuring the viscosity of the thick albumen. The higher the value, the fresher the albumen. This alternative method for measuring freshness combines the use of callipers and an egg diffusion chart. It is less accurate and a little trickier to undertake, and you will need to clean the callipers between each entry.

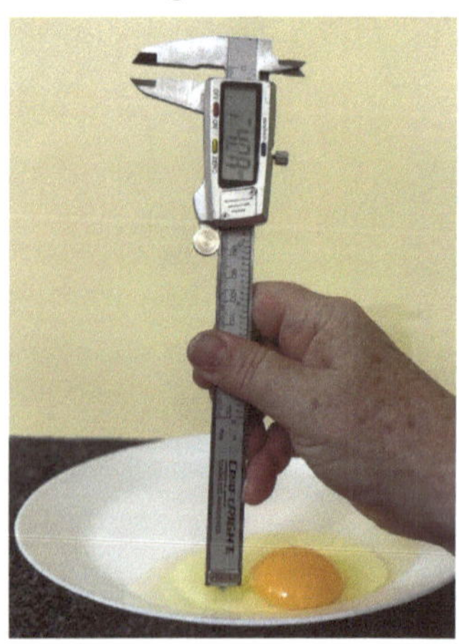

Callipers are used to measure the height of the albumen in millimetres.

Egg Diffusion Chart

Using the chart quickly identifies the furthest distance the **thick/ dense** albumen has spread from the centre of the yolk.

Having gathered the two measurements, use the following formula to calculate the albumen index ratio.

$$\frac{\text{Height of dense albumen (mm)}}{\text{Average diameter of albumen (mm)}} \times 100 = \text{Albumen Index}$$

Example

$$\frac{4.7\ \text{mm}}{30\ \text{mm}} \times 100 = 15.66$$

At the time of writing this book, there is no widespread use of a diffusion chart or callipers in Australia. It is difficult to get a reading if the plate is not perfectly flat on a level table. Instead, most judges use their experience to judge the freshness of the contents. Also, comparing entries to each other is another technique, just not as accurate.

Yolk

Colour

The colour of the yolk should be a rich golden yellow. It is recommended to use a yolk fan or colour chart if one is available (between 8 and 11 are considered golden).

This DSM-Firmenich colour card is shown with permission. These commercial fans are difficult to source and are expensive.

Specialised Breeders Australia is a company that raises and sells commercial chickens for both egg laying and meat production. They have kindly allowed me to show a picture of their promotional colour yolk card.

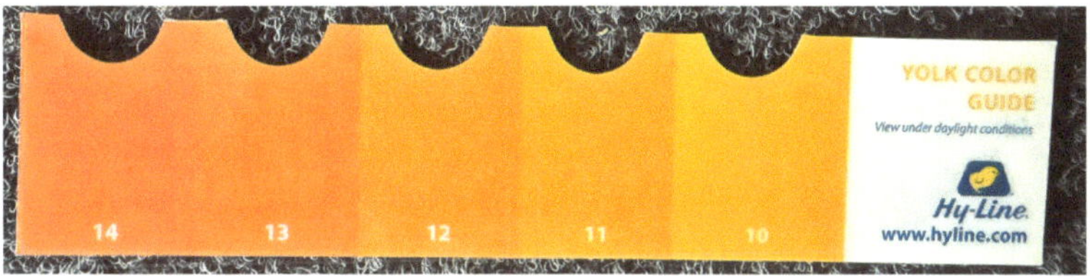

Eccentricity

Eccentricity is the position of the yolk within the albumen when the contents of the egg are on a flat surface. The measure ranges from one to 10 on the egg eccentricity grading chart, one being the ideal in the middle of the albumen, and 10 where the yolk is positioned partially outside the albumen.

What score would you give this egg? Refer to the chart on the next page.

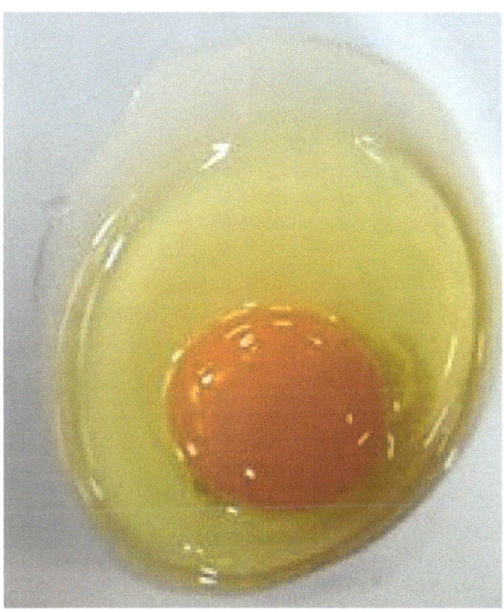

Egg Eccentricity Grading Chart

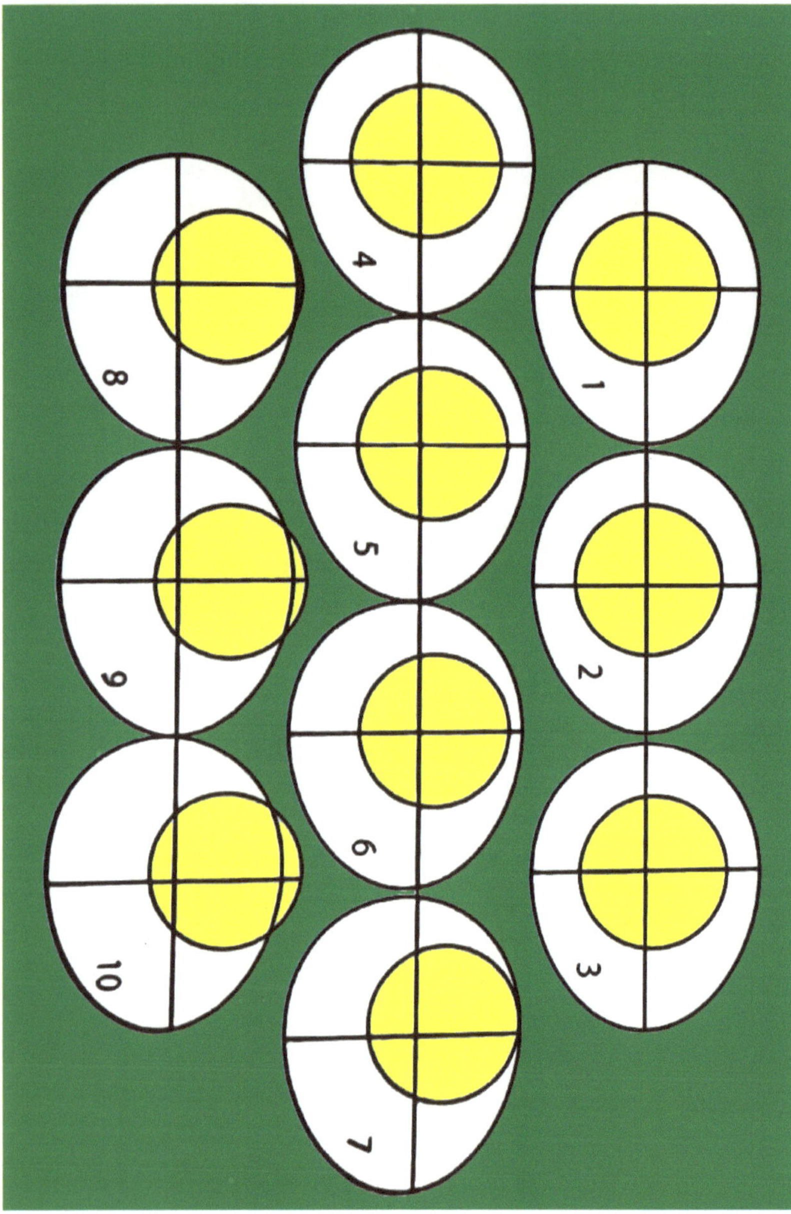

Mottling

Mottling is when the yolk has areas of colour differences across the surface. Mottling can be caused by:

- Diet (including certain weeds, raw soybeans, or high tannin content)
- High temperatures
- Stress (including heat stress)
- Certain medications like anti-coccidials
- Older age of hens
- Poor storage conditions

High storage temperatures or certain drugs necessitate discarding the egg. Yolks mottled from diet or age are generally safe to eat if the egg looks and smells normal. For exhibiting purposes, mottling is to be avoided.

Mottled	Good Even Pattern
	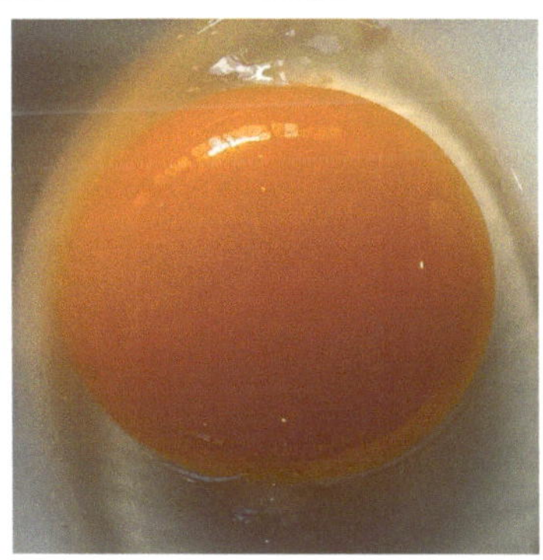

Blastoderm

Check if there are no signs of fertility, e.g. a "halo" around the germinal disc.

Check for Fertility

How to tell if an egg is fertile

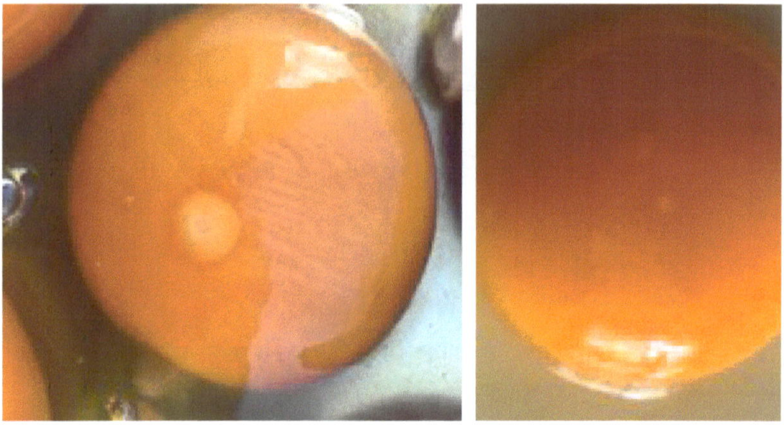

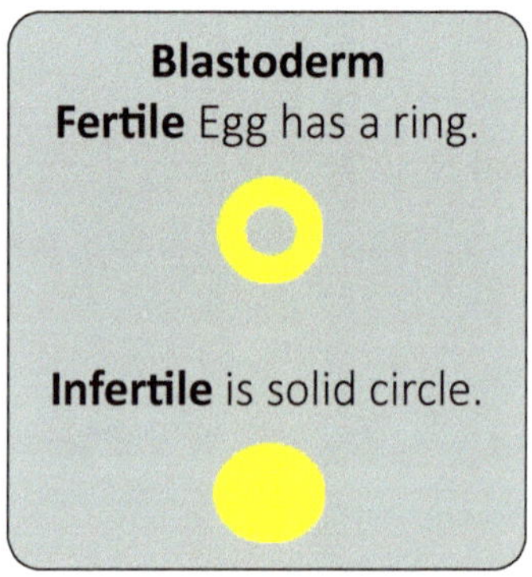

The examples shown are chicken eggs. It is not always as easy to see in other egg species, such as turkeys.

Serious Defect/Disqualification

Following is a list of reasons why eggs are passed over and not given a mark.

External

- Size too small or too large for class
- Weight — overweight in bantam, including contents classes.
- Shell colour pale or uneven
- The shell is dirty.
- Uneven shell structure: wrinkles, calcareous pimples, and

other deformations, such as lines or bulges.

- Soft/thin-shelled or porous
- Poor shape — spherical (round like a ball), narrow, oblong, asymmetrical.
- Excessive nest marks or scratches
- Uneven shell pattern
- Calcium pimples
- Over-prepared shell — clear evidence of cleaning, i.e., removed outer layer and lost bloom.

Internal

- Double yolks are considered a defect, although nice for eating. This is an issue for breeding as the egg is highly unlikely to successfully hatch and therefore be disqualified.
- Watery whites — where the viscosity of the albumen has reduced, and the albumen has spread across the plate. This occurs as the egg ages.
- Excessive blood streaks and meat spots in the albumen.
- Blood-stained yolks
- Fertile egg, with a developing embryo — "halo" around the blastoderm is considered a fault and will be disqualified.
- Artificial polish or colouring applied to the eggshell.

A prepared egg cannot be awarded "Egg of Show".

Example Defects

Double Yolk

Hole in Shell

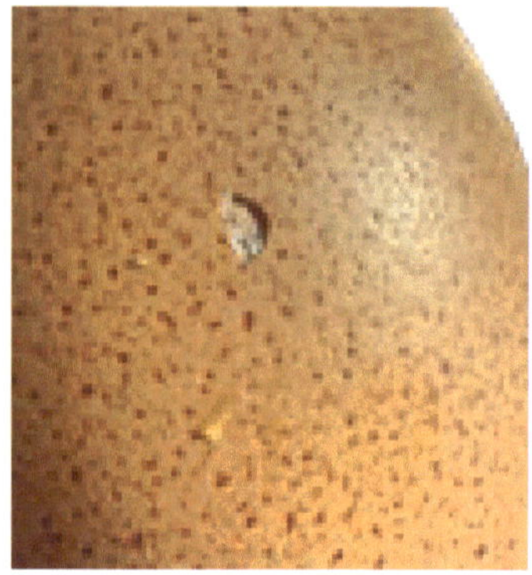

Wrinkled Shell

Artificial Colour

Prepared Egg Classes

Initially, the eggs should be judged for natural qualities such as shape, colour and matching eggs. These are some suggested criteria to judge the prepared aspect of the exhibit. The standard for the scale of points should be checked for the criteria on which to judge. It is recommended to keep a copy to hand from which to refer.

All Classes – Painted, Decorated, Displayed

Artistic Effect	Arranged decoratively and tastefully
Originality	Concept and the originality of an idea
Quality of Painting	Important for painted classes in particular, with allowances made for children's classes.
Use of Colour	Enhances the image for both painted and decorated eggs

In addition, for **decorated** eggs:

Quality of Construction	The egg may be mounted as part of the exhibit
Use of Materials	Use of different materials to enhance the effect of the display

In addition, for **displayed** eggs:

Quality and Matching	Eggs within the exhibit match in size and shape

Feedback to Exhibitors

When the judge takes the time to give feedback on an exhibit to the exhibitor, it can be viewed as a learning experience to assist them in improving their entries for subsequent competition.

This important aspect of judging should not be overlooked. I encourage Judges to provide constructive feedback in a sensitive manner, along with lots of encouragement.

Further Reading

Books

Title	Author	Publisher	ISBN
A Beginner's Guide to Backyard Poultry	James Finger	Bellsouth Publication	0-949340-00-6
Australian Poultry Standards 2nd Edition & Addendums	Various	Victorian Poultry Fanciers Association Ltd	978-1-921-48823-8
Backyard Poultry Medicine and Surgery - A Guide for Veterinary Practitioners	Cheryl B Greenacre, Teresa Y. Morishita	Wiley Blackwell	978-118-33543-7
British Poultry Standards 6th Edition	J. Ian H. Allonby, Philippe B. Wilson	Wiley Blackwell	978-1-119-50914-1
Chicken Health for Dummies	Julie Gauthier, Rob Ludlow	John Wiley & Sons Ltd	978-1-118-44427-6
Creative Poultry Breeding	Dr W.C. Carefoot	Veronica Mayhew	
Decorated Eggs	Ruby M Brown	Sally Milner Publishing	1-86351-181-4
Domestic Quail for Hobby or Profit	G.E.S. Robbins	World Pheasant Association	
Egg Decorating – Plain and Fancy	Arden J Newsome	Crown Publishers Inc	0-517-505967
Egg Nature's Miracle of Packaging	Robert Burton	William Collins Sons and Co Ltd	0-00-219437-6

Title	Author	Publisher	ISBN
Epic Eggs	Jennifer Sartell	Voyageur Press	978-0-7603-5222-9
Genetics of the Fowl	F.B. Hutt	Norman Creek Press	
Genetics in Practice	F.B. Hutt	Norman Creek Press	0-1938099079
Guidelines for Judging and Stewarding Exhibition Poultry	Multiple	Feather Clubs Association of Queensland Inc	N/A
Guinea Fowl Past & Present	Michael Roberts	The Gold Cockerel Series	0-947870-36-9
Keeping Geese Breeds and Management	Chris Ashton	The Crowood Press	978-1-84797-336-8
Keeping Quail A Guide to Domestic and Commercial Management	Katie Thear	Broad Leys Publishing	0-906137-26-8
Optimum Egg Quality – A Practical Approach	Jeffery A. Coutts, Graham C. Wilson	Department of Primary Industries and Fisheries, Queensland	0-9530150-6-8
Monographs of the Pheasant	Alain Hennache & Michael Ottaviani	World Pheasant Association	978-0-90686-454-8
Pheasants, including their care in the Aviary	H.A. Gerrits	Blanford Press	0-7137-06791
Poultry Eggs	Malcolm Thompson	Dampier Press Ltd	978-1-78926-681-8

Title	Author	Publisher	ISBN
Quail Past & Present	Michael Roberts	The Gold Cockerel Series	0-94787870-121
Selecting the Layers	W. Powell-Owen	Powell-Owen Poultry Publications	Out of Print
The Avian Egg	Alexis Romanoff & Anastasia Romanoff	Wiley	Out of Print
The Chicken Health Handbook - 2nd Edition	Gail Damerow	Storey Publishing	978-1-61212-013-3
The Illustrated Egg Handbook	Ayuko Kashimori	Context Publications	978-1-899-04374-3
The Most Perfect Thing. Inside (and outside) a Bird's Egg	Tim Birkhead	Bloomsbury	978-4088-5127-2
The New Incubation Book	A.F. Anderson Broen & G.E.S. Robbins	World Pheasant Assoc	0-86230-061-4
Turkey Production	Stanley J Marsden	Bellsouth Publication	0-949340-06-5

Online

There are many online resources available. These have not been included as they change too rapidly.

Index

Summary

This book is intended to be a quick reference to the basics of exhibiting eggs for the exhibitors. It is also intended to be a guide to running egg classes at shows for poultry clubs. The book also outlines the techniques of egg assessment in competition by judges.

The purpose of *Exhibiting Poultry Eggs* is to encourage good egg quality for both eating and breeding purposes, helping to promote good animal husbandry, and assisting in creating the healthiest of birds to produce the best eggs.

The judge is looking for consistency across all eggs within an entry, judging for compliance with the egg standard of the country. This book references the standards for Australia, New Zealand, the UK, Canada, and the United States. It is not limited to these standards, as it includes other potential species where eggs may be exhibited.

Poultry is defined as a domesticated bird. This book covers those species most kept for the purposes of meat and eggs. An egg can be exhibited from any poultry breed or variety, be it pure, commercial, or cross-bred.

Characteristics by which an egg is judged are determined by the description of the class. Therefore, it is important that those criteria of species, size, and colour are considered. It should be clearly stated in the show schedule as well as the rules/standards by which the eggs will be judged.

Prepared egg classes have become very popular and allow the opportunity for anyone to enter, even if they don't keep any poultry. Sometimes it becomes the first step into the hobby.

Classes for decorated and painted eggs can also be a big attraction at a show, providing a bit of fun and encouragement to exhibit eggs.

I hope you found this book useful.

www.ingramcontent.com/pod-product-compliance
Lightning Source LLC
LaVergne TN
LVHW051128210926
848482LV00001B/23
* 9 7 8 1 9 2 3 7 4 6 5 4 1 *